Climate Urbanism

"In times of a fast-changing world characterized by growing population, environmental change, biodiversity loss, contested economic structures, frequency of natural hazards and climate change, it is time to not only act but also ponder about possible futures in a planet of cities. This book is a timely addition, as a text that walks us through a soul-searching process on the blunt of climate change on cities and how city futures can be shaped. The authors highlight that there won't be a single vision for future cities but multiple futures and contextually based. The examples take us from climate concerns to how social systems, urban infrastructure and institutions can be reconfigured to design city futures that address climate threats that are just and inclusive in a new climate urbanism. The climate urbanism that this book examines in detail illustrates how important and critical cities are to respond to the climate urgency. That education, training and new professionalism is important for visioning and shaping future climate urbanism.

An academic and provocative book that is a good source for education, training and curriculum development and for readers interested in policy transformation for city futures that is socially transformative climate urbanism driven by social movements across cities in both Global North and South.

This is the material for city futures."

—Shuaib Lwasa, *Associate Professor, Department of Geography Geoinformatics and Climatic Sciences, Makerere University*

"With the outskirts of major cities still scorched by the devastating 2019–2020 fires season and major environmental issues in question at the heart of the COVID-19 crisis, there might be no better time to reform our climate approaches in cities. In *Climate Urbanism* Robin, While and Castán Broto prompt us to do so in a critical and progressive way. The volume seeks to make amend to the limits of climate-focused urban research, and to draw up a new agenda for action. In doing so they chart, collaboratively in dialogues with seventeen authors, what a research and action agenda for a 'climate urbanism' proper would look like. Rich in insights, both theoretical and empirical, emanating from a variety of viewpoints and experiences, from Colorado Springs and Maputo to Nepal via Malawi, from resilience and risk to the role of local governments in international policy via intergenerational learning, the volume is a truly refreshing mix of expertise with an explicitly transformative aim. Centrally,

Climate Urbanism does so as a new communal project asking us, academics, practitioners and even the broader public not to forget that urbanism is a broader undertaking that emergences from recognition of its multiple interpretations, and that there is no one single solution available out there. Progressive, emancipatory and explicit about its own contradiction, the book is just exactly the kind of urbanism it advocates, and one that is much needed at a time of profound crisis—really a must read for anyone itching to get out there and build a different kind of urban life."

—Prof Michele Acuto, *Director of the Connected Cities Lab, University of Melbourne*

"A critical and novel contribution from top climate social scientists examining the potential and contradictions of climate urbanism for an environmentally just and transformative development."

—Isabelle Anguelovski, *ICREA Research Professor and Director, Barcelona Lab for Urban Environmental Justice and Sustainability, Universitat Autònoma de Barcelona*

"*Climate Urbanism: Towards a Critical Research Agenda* provides a fresh lens through which to interpret the contours of urban change induced by living with and responding to climate change. The book's agenda is both generative and critical. By unpacking climate urbanism's underlying rationalities, contestations, knowledge frames, socio-material and politico-institutional forms, the book exposes the diversity of experiences from which new understandings of climate-related urban change must be developed. It offers no easy answers, but makes clear that producing just forms of climate urbanism will demand interdisciplinary knowledge that is both receptive to diverse urban experiences and acutely attuned to the politics inherent in climate-related urban change. The book poses challenging new and incisive questions of urban researchers and practitioners alike, while also suggesting productive new entry points for intervention."

—Prof Pauline McGuirk, *Director, Australian Centre for Culture, Environment, Society and Space (ACCESS), University of Wollongong NSW, Australia*

Vanesa Castán Broto
Enora Robin • Aidan While
Editors

Climate Urbanism

Towards a Critical Research Agenda

palgrave
macmillan

Editors
Vanesa Castán Broto
Urban Institute
University of Sheffield
Sheffield, UK

Enora Robin
Urban Institute
University of Sheffield
Sheffield, UK

Aidan While
Urban Institute
University of Sheffield
Sheffield, UK

ISBN 978-3-030-53388-5 ISBN 978-3-030-53386-1 (eBook)
https://doi.org/10.1007/978-3-030-53386-1

This Palgrave Macmillan imprint is published by the registered company Springer Nature Switzerland AG.
The registered company address is: Gewerbestrasse 11, 6330 Cham, Switzerland

Acknowledgements

This book emerges from the debates held at the international workshop "The New Climate Urbanism: exploring the changing relationship between cities and climate change" that took place at the University of Sheffield on September 4–5, 2019. As editors, we are enormously thankful to Victoria Simpson and Ryan Bellinson for their help in organising this event. We are also indebted to all the chapter authors who both presented their work at the workshop and then bought into our project to make an accessible book on a challenging topic. We would like to extend our thanks to all the contributors to the workshop, including those who were not able to participate in the final book. The debates at the workshop helped in structuring the book and all contributions benefitted from the conversations we had. We would also like to thank those who for different reasons could not attend the workshop at the last minute, especially Harini Nagendra. Our thanks go to Linda Westman, Ping Huang, Hita Unnikrishnan, and Erika Conchis for making our Climate Urbanism research theme what it is: a collegial, friendly, and stimulating research group committed to think about climate change and cities from new locations and disciplines. We are also enormously thankful to all our colleagues and visiting fellows at the Urban Institute who have created a vibrant intellectual atmosphere where our group on Climate Urbanism has thriven and grown: Simon Marvin, Rachel Macrorie, Beth Perry,

Bertie Russel, Jonathan Rutherford, Jonathan Silver, Victoria Habermehl, Sophie King, Andy Lockhart, Miguel Kanai, Michele Lancione, Abdoumaliq Simone, Pauline McGuirk, Sam Burgum, Tom Goodfellow, Lindsay Sawyer, Irit Katz, Rowland Atkinson, and Zarina Patel. Finally, we would like to thank the individuals in the Faculty of Social Sciences, University of Sheffield, who have made this project possible through their friendship and ongoing support: John Flint, Craig Watkins, Dorothea Kleine, Dan Brockington, Sarah Salway, Charlie Burns, and Nathan Hughes. This project was supported by the Leverhulme Trust (2016 Philip Leverhulme Prize on Geography).

Contents

1 Introduction: Climate Urbanism—Towards a Research Agenda 1
Vanesa Castán Broto, Enora Robin, and Aidan While
1.1 Introduction 1
1.2 What Is Climate Urbanism? 4
1.3 Climate Urbanism and Transformative Action 6
1.4 Knowing Climate Urbanism 7
1.5 Climate Urbanism as a New Communal Project 8
1.6 Conclusion 9
References 10

Part I What Is Climate Urbanism? 13

2 For a Minor Perspective on Climate Urbanism: Towards a Decolonial Research Praxis 15
Enora Robin, Linda Westman, and Vanesa Castán Broto
2.1 Introduction 15
2.2 Climate Urbanism as Just a Neo-colonial Project? 17

2.3 Postcolonial Thinking as a Way of Seeing, Decoloniality as a Research Praxis 19
2.4 Decolonizing Climate Urbanism 23
2.5 Conclusion 26
References 26

3 Climate Urbanism and the Implications for Climate Apartheid 31
Joshua Long, Jennifer L. Rice, and Anthony Levenda
3.1 Introduction 31
3.2 From Sustainable Urbanism to Climate Urbanism 32
3.3 Defining and Deconstructing Climate Urbanism 35
3.4 Against Climate Apartheid and Toward a Transformative Climate Urbanism 38
3.5 Conclusion 44
References 45

4 The New Climate Urbanism: Old Capitalism with Climate Characteristics 51
Linda Shi
4.1 Introduction 51
4.2 Metropolitan (Urban-Urban) Dynamics of Exclusionary Resilience 52
4.3 Territorial (Urban-Rural) Dynamics of Extractive Resilience 55
4.4 Conclusion: Imagining Alternative Climate Urbanisms 58
References 62

5 Understanding the Governance of a New Climate Urbanism 67
Sirkku Juhola
5.1 Introduction 67
5.2 The Institutionalization of Climate Policy 69
5.3 Implementing Urban Climate Policy 72

5.4 Policy Coherence and Competition 74
5.5 Conclusion 77
References 77

Part II Climate Urbanism and Transformative Action 81

6 Urban Climate Imaginaries and Climate Urbanism 83
Linda Westman and Vanesa Castán Broto
6.1 Introduction 83
6.2 Urban Climate Imaginaries 84
6.3 The Urban Within the International Climate Regime 86
6.4 Conclusion 91
References 92

7 Institutional Dynamics of Transformative Climate Urbanism: Remaking Rules in Messy Contexts 97
James J. Patterson
7.1 Introduction 97
7.2 Cities and Institutional Change 99
7.3 Institutional Dynamics of Urban Climate Transformations 101
7.4 Conclusions 108
References 109

8 Urban Resilience and the Politics of Development 117
Eric Chu
8.1 Introduction 117
8.2 Urban Resilience as a Contested Concept 118
8.3 Vignettes from Urban India 122
8.4 Conclusion: The End of Urban 'Resilience'? 128
References 131

9 Two Cheers for "Entrepreneurial Climate Urbanism" in the Conservative City 137
Corina McKendry
9.1 Introduction 137
9.2 Pro-growth City Environmentalism 139
9.3 Business Interests and the Closing of a Coal-Fired Power Plant 140
9.4 The Transformative Potential of Low-Carbon as a Growth Strategy? 143
9.5 Conclusion 145
References 146

Part III The Knowledge Politics of Climate Urbanism 151

10 An Adaptation Agenda for the New Climate Urbanism: Global Insights 153
Marta Olazabal
10.1 Introduction 153
10.2 Data and Methods 155
10.3 Results and Implications for Climate Urbanism 159
10.4 Conclusion 166
References 167

11 The New Climate Urbanism: A Physical, Social, and Behavioural Framework 171
Luna Khirfan
11.1 Introduction 171
11.2 Understanding Climate Urbanism Through Lynch's Framework 172
11.3 Situating Climate Urbanism(s): Lessons from Charlottetown, Amman, Negril, and Zürich 175
11.4 Conclusions: Efficiency and Justice in Climate Urbanism 188
References 190

12 Collaborative Education as a 'New (Urban) Civil Politics of Climate Change' 195
Andrew P. Kythreotis and Theresa G. Mercer
12.1 Introduction 195
12.2 Education, Citizen Engagement and the 'New Civil Politics of Climate Change' 197
12.3 Collaborative Education for Climate Change Education 200
12.4 Conclusion 205
References 206

Part IV Climate Urbanism as a New Communal Project 211

13 Community Energy Resilience for a New Climate Urbanism 213
Long Seng To
13.1 Introduction 213
13.2 Community Resilience and Climate Urbanism 214
13.3 Learning from Nepal and Malawi 217
13.4 Conclusions 222
References 223

14 Making Climate Urbanism from the Grassroots: Eco-communities, Experiments and Divergent Temporalities 227
Jenny Pickerill
14.1 Introduction 227
14.2 Conceptual Starting Points 228
14.3 Eco-communities 231
14.4 Self-built Infrastructures 232
14.5 Tensions of Self-Built Infrastructures 239
14.6 Conclusions 240
References 241

15 **Conclusions: Three Modalities for a New Climate Urbanism** 243
Vanesa Castán Broto, Enora Robin, and Aidan While
15.1 Introduction 243
15.2 Rethinking Multiple Pathways to the Future 246
15.3 Three Modalities of a New Climate Urbanism: Entrepreneurial, Reactive, Transformative 247
15.4 Conclusion 250
References 252

Notes on Contributors

Vanesa Castán Broto is a Professor of Climate Urbanism at the Urban Institute, University of Sheffield (UK). She leads the projects Low Carbon Action in Ordinary Cities, LOACT (European Research Council) and Community Energy and Sustainable Energy Transitions in Ethiopia, Malawi, Mozambique, CESET (UK Global Challenges Research Fund).

Eric Chu is an assistant professor in the Department of Human Ecology, University of California, Davis. He researches the politics of climate change governance in cities, with particular interest in socio-spatial change, development planning, and environmental justice. He has written extensively on social inclusion and equity in the context of climate adaptation in cities across the Global South.

Sirkku Juhola is Professor of Urban Environmental Policy at the Faculty of Biological and Environmental Sciences, University of Helsinki, where she leads the Urban Environmental Policy research group. She is also a guest professor at the Centre for Climate Science and Policy (CSPR) at Linköping University, Sweden. She has published widely on environmental and climate governance, vulnerability and adaptation to climate change. She is part of the Risk Knowledge Action Network (Risk KAN) Development Team and also a member of the Climate Panel of Finland, a scientific advisory body to the Government of Finland.

Luna Khirfan is Associate Professor at the School of Planning, the University of Waterloo (Ontario, Canada). Her work underscores local communities' engagement in urban planning and design specifically adapting built form to climate change. She transforms the conventional "participatory design" role of the charrette into a method of data collection and mechanism of knowledge exchange with local communities. She also investigates daylighting urban streams for climate change adaptation and mitigation. She is Lead Author for Chapter 6: "Cities, settlements and key infrastructure" as part of the Working Group II contribution to the upcoming IPCC Sixth Assessment Report.

Andrew P. Kythreotis is Senior Lecturer in Social and Political Geography at the School of Geography, University of Lincoln, UK, a senior researcher at the Tyndall Centre for Climate Change Research, University of East Anglia and an honorary research fellow at the School of Psychology, Cardiff University. His research and teaching revolve around the broad themes of climate change and the environment and how its policy, politics and governance are constructed around socio-spatial ontologies. He has advised national governments on climate change adaptation issues, has recently co-founded the Lincoln Climate Commission and was an external reviewer for the 2017 and upcoming 2022 DEFRA's UK Climate Change Risk Assessment Evidence Reports.

Anthony Levenda is an assistant professor in the Department of Geography and Environmental Sustainability at the University of Oklahoma (USA). His work focuses on the politics and governance of environmental change, especially related to energy and cities. Current research explores (1) how diverse groups frame and implement just energy transitions and (2) the urban politics of climate mitigation and adaptation.

Joshua Long is Associate Professor of Environmental Studies and Chair of the Environmental Studies Program at Southwestern University (USA). His research interests include climate justice, critical urban sustainability, environmental justice, and ontologies of nature. His recent research is focused on the repercussions of solutions and political responses to the climate crisis.

Corina McKendry is Associate Professor of Political Science and Director of the Environmental Studies Program at Colorado College, a liberal arts college in Colorado Springs (USA). She teaches and writes on subnational climate governance in the Global North, with a particular focus on the relationship between equity, environmental effectiveness, and economic growth in city greening initiatives. Other work focuses on environmental politics in the Rocky Mountain West and broad questions of subnational sovereignty and energy democracy in the context of climate change. She received her PhD in Politics from the University of California, Santa Cruz.

Theresa G. Mercer is Senior Lecturer in Biogeography and Planetary Health at the School of Geography, University of Lincoln (UK). She has held several academic postings at the University of Hull, UK; Cranfield University, UK; Cardiff University, UK; Keele University, UK and the University of Queensland, Australia. She is an interdisciplinary environmental scientist with broad interests in environmental toxicology, environmental management, and Education for Sustainable Development.

Marta Olazabal is a research fellow and head of the Adaptation Research Group at the Basque Centre for Climate Change, BC3. She has a background in environmental engineering and a PhD in land economy. She researches urban sustainability and climate change governance, the processes of creation and application of usable knowledge in systems thinking and in approaches that bridge science and policy. Her current research focuses on understanding the ways in which adaptation might be enabled in cities worldwide. She is funded by AXA Research Fund and by the Spanish government.

James J. Patterson is Assistant Professor of Institutional Dynamics in Sustainability at Utrecht University, The Netherlands. He has a broad cross-disciplinary perspective situated at the intersection of environmental studies and political science. His research focuses on the institutional and political dynamics of addressing complex collective problems. For example, how and why do institutions change, and with what effects on governance systems and society? The main substantive focus of his work is climate change governance in the domestic political sphere.

Jenny Pickerill is Professor of Environmental Geography at the University of Sheffield. Her research focuses on inspiring grassroots solutions to environmental problems and in hopeful and positive ways in which we can change social practices. She has published three books (*Cyberprotest; Anti-war Activism; Eco-Homes*) and over 30 articles on themes around environmentalism, Indigenous geographies, anarchism and eco-housing. Information about her work and contact details are at www.jennypickerill.info

Jennifer L. Rice is an associate professor in the Department of Geography at the University of Georgia (USA). She is an affiliate faculty with UGA's Institute for Women's Studies and Center for Integrative Conservation Research. Her research interests include: urban political ecology, climate and carbon governance, nature-society theory, scholar-activism, feminist and anti-racist praxis and pedagogy. She has conducted research on these issues in Seattle, Washington, and southern Appalachia. Her most recent research is focused on the intersections of climate and housing justice activism in Seattle. She is also the co-convener of the Athena Co-Learning Collective (AthenaCollective.org).

Enora Robin is a research fellow at the Urban Institute (University of Sheffield) where her work explores contemporary urban transformations across three interrelated themes, namely (1) the politics of urban expertise, (2) everyday climate action, and (3) infrastructural change. Her research focuses on the role of low-income communities in the provision of decentralised and community-owned energy systems in Ghana and Mozambique.

Long Seng To is an engineer for Development Research Fellow funded by the Royal Academy of Engineering at Loughborough University (UK). Her research tackles the challenge of providing access to affordable, reliable, sustainable, and modern energy for all in the context of increasing stresses and shocks, such as climate change, disasters, and conflicts. Her research focuses on enhancing community energy resilience using renewable energy in South Asia and sub-Saharan Africa.

Linda Shi is an assistant professor at Cornell University's Department of City and Regional Planning (USA). She studies how cities' efforts to adapt to climate impacts contribute to social and spatial inequality, how urban land governance institutions shape climate vulnerability and adaptation responses, and the potential for alternative property rights regimes and governance models to enable more equitable and sustainable outcomes. She comes at these issues having worked on watershed restoration, water and sanitation, and development planning in the US, Latin America, Asia, and Africa. She has degrees in urban planning and environmental management from Yale, Harvard, and MIT.

Linda Westman is a research associate at the Urban Institute, University of Sheffield (UK). Her research engages with climate governance, transformative change, and justice. Her work explored the politics of climate action in urban China, which involved re-imagining the concepts of multilevel governance, partnerships, and governing rationalities from the perspective of China's political system. Her recent publication (*Urban Sustainability and Justice*, with Vanesa Castán Broto) examines postcolonial and feminist critiques of urban sustainability discourses. Prior to her current position, she was a postdoctoral researcher at the University of Waterloo and a policy analyst at the Embassy of Sweden in Beijing.

Aidan While is a senior lecturer in the Department of Urban Studies and Planning and co-director of the Urban Institute at the University of Sheffield. He has researched urban environmental policy since the early 1990s, charting the ebb and flow of climate policy and its intersections with economic and social policy in cities in Europe, Asia, and North America.

List of Figures

Fig. 6.1	Constitution of urban climate imaginaries and their materialization in narratives of action, policies, and situated experiences that shape both the international climate regime and the local context of action	86
Fig. 8.1	Large-scale flood protection infrastructure along the Tapi River in Surat. (Photo taken by author)	124
Fig. 8.2	View from Fort Cochin towards Vypin and the Vallarpadam International Container Terminal. (Photo taken by author)	127
Fig. 10.1	Documenting protocol (Olazabal, de Gopegui, et al. 2019)	156
Fig. 10.2	Number of adaptation policies at national, regional and local levels per city (Olazabal, de Gopegui, et al. 2019). Orange bubbles indicate the total number of plans; white bubbles indicate the total number of local (city and metropolitan) plans. Sampled countries are shaded gradually indicating the number of national policies	160
Fig. 11.1	A composite showing the potential of fit's manipulability and reversibility and vitality's sustenance, consonance, and safety to alleviate Charlottetown's climate-related risks. (Diagrams: designed by the author and executed by Anna Maria Levytska and Rachel Rauser)	176
Fig. 11.2	Severe beach erosion in Negril, Jamaica, triggered by sea-level rise and extreme weather events	180

Fig. 11.3	Sandals Negril Beach Resort installed artificial reef and buffered parts of its shoreline with rocks	181
Fig. 11.4	Small hotels in Negril used sandbags and installed impromptu structures to address severe beach erosion and keep the seawater at bay	182
Fig. 11.5	Rainwater harvesting tanks at Negril's Rockhouse Hotel	184
Fig. 11.6	Attempts by locals to regenerate the mangroves in Orange Bay, Negril	184
Fig. 11.7	Local activism led to the protection of Negril's Great Morass from development (Map drawn by Tapan K. Dhar)	186
Fig. 11.8	The cooperation between urban developers and municipal authorities in Zürich in implementing the Bachkonzept reflects the market's creative continuism while also transforming the urban form in unique ways	187
Fig. 11.9	A composite showing the positive impact of Zürich's daylighting policy as a nature-based solution that simultaneously provides vital ecosystem services and urban spaces for education, recreation, and leisure	188
Fig. 13.1	A conceptual framework for community energy resilience (To and Subedi 2020)	217
Fig. 14.1	Green Hills eco-community	233
Fig. 14.2a and 14.2b	Green Hills spring water and rainwater systems	234
Fig. 14.3	LILAC physical and social infrastructures	238

List of Tables

Table 2.1	Principles for a minor perspective on climate urbanism	24
Table 7.1	Types of institutional dynamics in urban climate transformations	101
Table 10.1	Number of countries and cities per world region with no planning (Olazabal, de Gopegui, et al. 2019)	159

1

Introduction: Climate Urbanism—Towards a Research Agenda

Vanesa Castán Broto, Enora Robin, and Aidan While

1.1 Introduction

This book argues that as climate change dramatically reshapes how we understand, imagine, live, and intervene in cities, a New Climate Urbanism is emerging as a way to rethink and reorient urban life. This emergence is rooted in the decade-long recognition that significant actions on cities and urbanization are required as crucial elements in policy responses to climate change (Bulkeley 2013). Rapid urbanization is a major driver of climate change and cities have become important sites for adaptation and mitigation efforts and for developing climate-resilient development pathways (Revi et al. 2014; IPCC 2018). Around the world, urban areas have to respond to the new realities created by a changing climate, including enhanced exposure to climate risks and climate-induced migration. These challenge existing forms of urban management,

V. Castán Broto (✉) • E. Robin • A. While
Urban Institute, University of Sheffield, Sheffield, UK
e-mail: v.castanbroto@sheffield.ac.uk; e.robin@sheffield.ac.uk; a.while@sheffield.ac.uk

V. Castán Broto et al. (eds.), *Climate Urbanism*,
https://doi.org/10.1007/978-3-030-53386-1_1

particularly at a time when cities are adjusting to a new context for urban public health management following the COVID-19 pandemic. With public health and climate crises increasingly overlapping, urban futures now look less certain and more insecure. Some are hoping for the COVID-19 pandemic to be the opportunity to rethink the kind of system and relations that brought us here and to catalyse transformations for promoting long-term sustainability.

Debates around climate change politics, governance, and vulnerability have thus elevated discourses of urban transformation: climate change politics has become an essential driver of new models of urbanism (Castán Broto 2017). Increasing protection against environmental threats is a priority for climate-proofing and retrofitting efforts (Bulkeley et al. 2015). Defensive projects, such as large infrastructural projects and the securitization of ecological enclaves are enhancing existing and creating new inequalities (Hodson and Marvin 2009; Rice et al. 2020). These new models constitute a New 'Climate Urbanism' (Long and Rice 2019), whereby different actors within and across urban areas re-define what cities ought to be in a changing climate. This edited volume is the first attempt to map, systematically, the contours of a research agenda on this New Climate Urbanism. Our focus is not only on the material and technical elements that make up urban spaces but also on the changing social practices that follow major transformations of urban cultures. We use the term 'urbanism' because, as geographer Eugene McCann (McCann 2017) explained, we are looking into how urban life is defined and understood at different points in time (and space), and how this has historically shaped different ways of intervening in cities. Today we find ourselves in an intensified 'climate moment' for cities, as climate change transforms both how we live in urban areas and how we govern them in fundamental ways. This edited volume sets out to explore the challenges posed by the emergence of this new paradigm, starting with the question: *what does climate urbanism consist of, and how does it differ from other models of urbanism?* This question articulated discussions during a two-day workshop on the *New Climate Urbanism*, hosted by the Urban Institute (University of Sheffield) in September 2019, which brought together many of this book's contributing authors. Following from this meeting, this book aims to develop a research agenda on the constitutive elements of climate urbanism, its drivers, and its impacts. The contributions gathered here examine the rationalities underpinning how climate urbanism is embraced, promoted, or contested, and how it transforms socio-material fabric of cities, addressing one or more of the following research questions:

- How can we define climate urbanism?
- What type of expertise and knowledges are produced, mobilized, and needed in this new era of climate urbanism?
- What are the absences and silences in research on cities and climate change? What should be the research priorities for the future?
- How can a research agenda on climate urbanism encompass the diversities of planetary urban conditions?
- How can researchers engage with climate urbanism to make a difference to policy and practice, to create and deliver environmentally just transformations?

This book offers some preliminary takes on these debates, bringing together thirteen contributions from a range of scholars in the field. These contributions aim not only to map the whole gamut of possible research directions on climate urbanism but also to foster multi-disciplinary dialogue. Adapting and responding to climate change will require solutions that mobilize different epistemological and theoretical perspectives in different geographical locations. In this book, we wish to emphasize the importance of accounting for differences in theorizing urban life under climate change—beyond the experiences of cities of Europe and North America. We also call for an honest reckoning with the limits of our disciplinary knowledge to understand the manifold ways in which urban areas change in the age of climate change. All the contributions were written before the COVID-19 crisis. However, climate change is inherently an issue of human health and well-being. The pandemic has therefore not changed our intent: rather, it has reaffirmed the need for a concept of human protection that engages with the urban collective. A safe city is one that addresses both climate change impacts and public health risks in a just manner, and this requires a rethinking of how we, as human collectives living in cities, relate to nature within and beyond our cities.

We have divided this book into four parts that help us conceptualize and trace the contours of a research agenda on climate urbanism.

- Part I asks 'What is climate urbanism?' and explores the key features of climate urbanism from different locations and epistemological traditions, highlighting the shortfalls of dominant theorizations of climate urbanism, firmly grounded in research in North America;
- Part II develops a critical perspective on the transformative potential of climate urbanism, particularly its ability to challenge social and environmental injustices;
- Part III focuses on climate urbanism as a knowledge-mobilizing process. It links knowledge production to the delivery of climate urbanism as a distinctive mode of urban development and critically interrogates current knowledge paradigms underpinning climate and urban science;
- Part IV envisages the delivery of climate urbanism as a new communal project, focusing on the role of citizens and non-state actors in driving transformative climate urbanism. It seeks to broaden the definition of urbanism beyond a focus on state and private actors to identify more radical pathways for the implementation of just climate action in cities.

1.2 What Is Climate Urbanism?

The book aims to define what climate urbanism is, and the extent to which it differs from previous theorizations of the relationship between climate change and urban transformations. Rather than emerging as a consistent model for urban development or as a compendium of characteristics in the city, a research agenda on climate urbanism has emerged as a critique of how climate change is addressed in contemporary cities. In Chap. 2, Enora Robin, Linda Westman, and Vanesa Castán Broto call for a minor theory of climate urbanism, arguing that existing research has failed to develop theorizations of climate urbanism that reflect the diversity of urban conditions. In doing so, the authors set out future 'research-praxis' directions that articulate minor perspectives into broader theorizations of climate urbanism. In Chap. 3, Joshua Long, Jennifer L. Rice, and Anthony Levenda extend previous arguments about climate urbanism as an approach characterized by the emergence of new governance arrangements centred around carbon control and securitization (see also Rice 2010; Rice et al. 2020). To them, these approaches to climate urbanism create new logics of climate apartheid, furthering and creating new form of inequalities in cities.

Scholarly engagement with climate urbanism requires to critically explore its current manifestations and to expose its most damaging impacts, but it also implies exploring how it could be appropriated as a progressive tool to reimagine urban life in the age of climate change. Following Long, Rice, and Levenda, most chapters in this book approach climate urbanism as a polyvocal and generative concept that can be mobilized to engage with pressing issues such as climate-related segregation (or 'climate apartheid') and to rethink the relationship between urbanization, cities, and broader processes of ecological violence and dispossession. In this vein, in Chap. 4, Linda Shi stresses two limitations of climate urbanism as currently framed. Firstly, climate change forces us to think through connexions that extend beyond the city's limits, and future research should not solely focus on cities as a unit of analysis. Secondly, the legacy of already well-known structures of oppression (e.g., colonialism, capitalism) should also be recognized. To her, the emerging framing of climate urbanism is more a manifestation of late capitalist urbanization with climate characteristics than something specifically new. This provocation invites us to consider whether an emerging research agenda on climate urbanism is likely to generate novel insights to support pathways for transformative actions in cities. In Chap. 5, Sirkku Juhola develops a theoretical framework to how climate challenges give rise to new ways of governing cities. This proposition addresses Shi's concerns about the novelty of the 'New' Climate Urbanism, paving the way for a systematic exploration of how urban governance is reconfigured as a result of climate change. Overall, what we call a New Climate Urbanism is distinctive insofar as it enables the analysis of a significant qualitative shift in the way we think about and act in cities under climate change. Still, many of the contradictions embedded in current forms of urbanism remain entrenched in the way urban areas are approached and understood in the context of climate change.

1.3 Climate Urbanism and Transformative Action

The second part of the book engages with a deliberate concern for the ability of climate urbanism to foster just urban transformations. In Chap. 6, Linda Westman and Vanesa Castán Broto show that cities and local governments are still ignored as transformative agents in international climate policies. Working through these tensions and speaking to the question of scale, a central issue for climate urbanism research will be its capacity to reframe the relationship between national and local governments. Climate urbanism may be a mechanism that reinforces urban governance to enable responses to climate change. In Chap. 7, James J. Patterson brings to the fore the question of institutional change to conceptualize transformation towards progressive forms of climate urbanism. In doing so, he stresses the need for future research to understand processes of change within historically and socially distinct settings. In Chap. 8, Eric Chu considers the concept of urban resilience and the way it shapes urban development strategies to address climate change, drawing on the experiences of two Indian cities. His work stresses how popular 'climate-friendly' concepts can spur or hinder transformative action on the ground. In Chap. 9, Corina McKendry explores the integration of climate action into the growth agenda of Colorado Springs, a conservative US city led by a climate-denying mayor. Her provoking intervention shows that even the most conservative cities can implement climate-friendly strategies when it suits their economic interests. In this example, social justice and climate change are not political arguments that local leaders put forward to justify low-carbon investments, even when some of those benefit low-income communities. The example stresses the importance of rethinking the geography of climate urbanism to move away from cities that portray themselves as climate leaders and to consider a range of experiences where climate action is implemented out of economic necessity rather than political commitment to address climate change. Overall, there is no evidence yet that climate urbanism is in any way transformative. Defining what is transformative and the language of transformation are themselves questionable. Perhaps we should focus more on rethinking climate change politics in the city and analysing urban change, rather than trying to define what transformations should look like a priori, as they become a new unicorn in urban theory.

1.4 Knowing Climate Urbanism

Issues of knowledge politics are central to climate urbanism. Expanding the geography of climate urbanism research requires to look at the political economy of climate and urban knowledge. At the moment, there have been some efforts to explore how knowledge politics shapes different models of climate urbanism. In Chap. 10, Marta Olazabal examines local governments' capacity to address climate change through knowledge of risks and adaptation strategies. Her review highlights that local adaptation plans are still lacking in many cities around the world. Climate adaptation plans, when they exist, often lack evidence on climate risks. The implications of her research are enormous. If adaptation plans are not built on evidence, what are they built on and what purposes do they meet, beyond displaying local governments' climate concerns?

In Chap. 11, Luna Khirfan emphasizes the multiple relationships that co-constitute climate urbanism. Her contribution shows how design-centered knowledges can help understand how the socio-physical and behavioural components of urban systems are reconfigured through environmental change. Her contribution establishes a counterpoint to other critical contributions in this book by highlighting the potential of design thinking to support climate adapted urban futures. Strengthening collaborations between socio-ecological, institutional, and geographical research with more practice-oriented disciplines, such as urban design, architecture, and planning, will be essential for climate urbanism research to shape urban trajectories going forward. Furthering partnerships with state, private, and civic actors, among which citizens, will also be essential to support transformative actions on the ground. In Chap. 12, Andrew

P. Kythreotis and Theresa G. Mercer explore the potential of new educational strategies as knowledge production processes that support intergenerational learning and empowerment. Their work raises questions as to how future research on climate urbanism can integrate innovative pedagogical practices and new forms of collaborations. Overall, the critical questions raised by the multiple expressions of climate urbanism in contemporary cities are intrinsically linked to the processes of knowledge production that make them possible. Reimagining more just forms of climate urbanism requires examining alternatives to the hegemonic knowledges that dominate planning and management in contemporary cities.

1.5 Climate Urbanism as a New Communal Project

Are there any alternative approaches to climate urbanism that prevent deepening urban inequalities? We generally celebrate forms of community-based action based on collective solidarities. However, the evidence of community-driven projects challenging dominant modes of climate urbanism is patchy, at best. The final part of the book brings together two contributions focusing on the role of communities in producing socially just and transformative climate urbanism. In Chap. 13, Long Seng To draws on the example of community projects in Nepal and Malawi as useful forms of governance to support adaptation to climate risks. She argues that community-led energy projects are more attuned to the specificity of local hazards and exposure to risks and build on local knowledge(s) in the context of decentralized governance. In Chap. 14, Jenny Pickerill reflects upon her long experience of studying eco-communities, showing that those provide opportunities to change broader cultures of relating to nature, environment, and resource flows. She also stresses the inherent contradictions built into eco-communities and how those limit their potential for larger transformative action. Both examples highlight that even in reimagining new communal responses to climate urbanism there is not a ready-made solution to sustain a just city under climate change.

1.6 Conclusion

The contributions brought together in this book suggest the emergence of different modalities of climate urbanism, which can be progressive and emancipatory (To; Pickerill; Kythreotis and Mercer) but never without contradictions (Rice, Long and Levenda; Patterson; Shi; Chu; McKendry). Different modalities of climate urbanism coexist and, at times, conflict with each other as their emergence is shaped by differing historical, socio-ecological, cultural, and political processes (Robin, Westman and Castán Broto).

Read together, these chapters showcase multiple forms of climate urbanism that we can group into three main modalities: reactive, entrepreneurial, or transformative. Reactive climate urbanism relates to the actions taken in cities to simply deal with the noticeable impacts of climate change. Indeed, violent physical and ecological transformations are already taking place, and cities are bearing the brunt of their impacts: from heatwaves to flooding, rising sea levels to melting glaciers, and dwindling water reservoirs, the resource security challenge has become central to the governance of cities. For example, in February 2018, the city of Cape Town was the first to introduce water restrictions after a drought depleted its reservoirs. Other cities like Maputo found themselves in similar situations, with citizens turning to private solutions to deal with water shortages. There is also growing evidence that climate change fosters large population displacements towards small and medium cities. The International Displacement Monitoring Centre has argued that 18 million people were displaced in 2017 due to climate change-related disasters. Cities are coping, for better or worse, with these enormous changes. Sometimes the consequences are born by the poorer sectors of the population, further increasing urban inequalities and vulnerability to new risks, as explored throughout this book. Entrepreneurial climate urbanism relates to a trend in climate urbanism, whereby climate change is seen as a new opportunity to foster economic competitiveness in cities, and where climate change impact worsens inequalities and competition between cities. The emphasis on 'entrepreneurial' aims to highjack the discourses of those proponents of uncritical optimism, which see this form of climate change-oriented disaster capitalism as a new wonderland for opportunities. In this way, the existential challenge posed by climate change is appropriated to reimagine current capitalist systems: under climate change, cities are set to develop a green, circular economy that will follow with a bountiful of economic benefits and jobs. Finally, transformative climate urbanism refers to the growing efforts by multiple actors to use cities as platforms for a broader transformation through different modes of experimentation with technologies and social life, or through insurgent forms of activism that lead to broader social mobilization. These efforts to implement climate action represent, in essence, an attempt to reconfigure the boundaries of what is acceptable and desirable. Identifying the drivers of change to transition to a particular modality of 'climate urbanism' will be a fundamental task for future research on the topic, and it closely relates to the question of scale, as discussed by many contributors in this book.

We will return to these modalities in the conclusion of the book, where we will evaluate their potential in terms of generating both a research agenda and an agenda for action on climate urbanism going forward. Transformative climate urbanism holds the most potential for environmentally just forms of urban development, but the contradictions embedded in urban climate action are also evident. Moreover, the pathways to transformative action will vary across geographies. This edited book offers some starting points for researchers and practitioners to think through the current and future reconfiguration of life in cities in a climate-changed world without losing sight of the diversity of urban life.

References

Bulkeley, H. (2013). *Cities and climate change*. Abingdon: Routledge.

Bulkeley, H., Castán Broto, V., & Edwards, G.A.S. (2015). An urban politics of climate change: Experimentation and the governing of socio-technical transitions. Abingdon, NY: Routledge.

Castán Broto, V. (2017). Urban governance and the politics of climate change. *World Development, 93*, 1–15.

Hodson, M., & Marvin, S. (2009). "Urban ecological security": A new urban paradigm? *International Journal of Urban and Regional Research, 33*(1), 193–215.

IPCC. (2018). Global Warming of 1.5°C. An IPCC Special Report on the impacts of global warming of 1.5°C above pre-industrial levels and related global greenhouse gas emission pathways, in the context of strengthening the global response to the threat of climate change, sustainable development, and efforts to eradicate poverty [Masson-Delmotte, V., P. Zhai, H.-O. Pörtner, D. Roberts, J. Skea, P.R. Shukla, A. Pirani, W. Moufouma-Okia, C. Péan, R. Pidcock, S. Connors, J.B.R. Matthews, Y. Chen, X. Zhou, M.I. Gomis, E. Lonnoy, T. Maycock, M. Tignor, and T. Waterfield (Eds.)].

Long, J., & Rice, J. L. (2019). From sustainable urbanism to climate urbanism. *Urban Studies, 56*(5), 992–1008.

McCann, E. (2017). Governing urbanism: Urban governance studies 1.0, 2.0 and beyond. *Urban Studies, 54*(2), 312–326.

Revi, A., et al. (2014). Urban areas (Chapter 8). In: Climate Change 2014: Impacts, adaptation, and vulnerability. Part A: Global and sectoral aspects. Contribution of Working Group II to the Fifth Assessment Report of the Intergovernmental Panel on Climate Change.

Rice, J. L. (2010). Climate, carbon, and territory: Greenhouse gas mitigation in Seattle, Washington. *Annals of the Association of American Geographers, 100*(4), 929–937.

Rice, J. L., Cohen, D. A., Long, J., & Jurjevich, J. R. (2020). Contradictions of the climate-friendly city: New perspectives on eco-gentrification and housing justice. *International Journal of Urban and Regional Research, 44*(1), 145–165.

Part I

What Is Climate Urbanism?

2

For a Minor Perspective on Climate Urbanism: Towards a Decolonial Research Praxis

Enora Robin, Linda Westman, and Vanesa Castán Broto

2.1 Introduction

In her now mythical essay on minor theory, Cindi Katz argued that "*the minor is not a theory of the margins, but a different way of working with material*" (Katz 1996: 489). In making this claim, she stressed the struggle involved in making knowledge claims from the margin. In her essay, there is a sense of frustration with the fact that queer and feminist theorists have long described the struggles of those who are 'minorized' and have demonstrated once again how racialized histories of colonialism shape both current affairs and dominant ways of thinking. In response to this hegemony, she argued that a minor theory would support the recognition of new subjectivities and agency for those who are labelled as 'other,' the 'minors'—the female, the postcolonial, the racialized, the

E. Robin (✉) • L. Westman • V. Castán Broto
Urban Institute, University of Sheffield, Sheffield, UK
e-mail: e.robin@sheffield.ac.uk; l.westman@sheffield.ac.uk; v.castanbroto@sheffield.ac.uk

V. Castán Broto et al. (eds.), *Climate Urbanism*,
https://doi.org/10.1007/978-3-030-53386-1_2

indeterminate, and the undesired. A minor theory offers a promise of change through the revelation of relationships of oppression, but to her it still found obstacles in altering dominant ways of producing knowledge.

In this chapter—written more than twenty years later—we share Katz's frustration and take her essay as a point of departure to ask what a 'minor' perspective on climate urbanism might be. In asking this question, we hope to address our own dissatisfaction with current debates on climate urbanism. We are frustrated with the (limited) attempts to develop alternative ways of understanding why and how climate urbanism matters, not from a privileged vantage position but rather from a minor perspective. For instance, research to date has not fully addressed the experiences of postcolonial cities, and postcolonial and decolonial theories have not yet been integrated into intellectual debates on climate urbanism (Robin and Castán Broto forthcoming). Furthermore, we struggle to see how climate urbanism research can become part of the broader response to the planetary climate crisis. This frustration forces us to ask if and how our scholarship can contribute to addressing the many injustices causing and reinforced by climate change. It also invites us to collectively engage in "working and reworking theoretical productions from the inside" (Katz 1996: 497). There is an obvious need to develop a minor perspective on the urban as a site of intervention in a climate-changed world. Yet, how can we establish such perspective if the topic of climate urbanism is still not attracting a critical mass of thinkers who are both researching and thinking from marginalized vantage points, particularly from the postcolony? Relatedly, what legitimacy do we have to develop this minor viewpoint as three white women trained in Europe? We find inspiration in postcolonial and decolonial scholarship, but how far can we claim it as ours? To what extent can we account for and address intersecting experiences of racialization, gender-based violence, and exclusion in climate urbanism research? How can we bring into our analyses the knowledge of the minor, without making it 'other,' while recognizing it as part of the context in which climate urbanism emerges? These questions are not merely rhetorical. They aim to highlight our own positions in the (academic) world and to stress that there are unavoidable absences and oversights in the perspective we present here.

In this opening chapter, we draw on postcolonial theory and decolonial thinking to explore what a minor perspective on climate urbanism might look like. This chapter however does not offer a definite 'minor theorization' of climate urbanism. Rather, it should be read as an invitation to expand the community of scholars involved in researching this topic to understand both its limits and emancipatory potential. What would a postcolonial perspective on climate urbanism entail? How could it be deployed? This is not an agenda for which we have ready-made conclusions, but one which is only beginning to unfold. In what follows, we first unpack current theorizations of climate urbanism and analyse their shortfalls. We then explore what a minor perspective on climate urbanism could look like, drawing on postcolonial theory and decolonial thinking and praxis. We acknowledge the necessary incompleteness of our proposition, but see it as an invitation to start looking collectively for alternatives to address one of the major challenges of our times.

2.2 Climate Urbanism as Just a Neo-colonial Project?

Critical urban scholarship has shown that how we understand and respond to climate change is a fundamentally political project (Luque-Ayala et al. 2018). Climate urbanism relates both to the localization of climate politics at the city level and to new ways of intervening in cities as a result of climate change (see Chap. 1). It points towards the emergence of new programmatic visions of the urban and the mobilization of substantial financial resources to implement those. From international organizations to city networks, philanthropic organizations, real estate federations, IT companies, and research institutes advising on urban climate policies, powerful coalitions have emerged to encourage public and private sector investments into climate-proof and smart urban projects (e.g. Goh 2019a; Rapoport 2014; Rashidi et al. 2019). Such processes are not new. Earlier iterations of the 'sustainable' and 'smart' city alongside increased fragmentation of infrastructures and the securitization of urban space have long been central to the management of environmental change, and contributed to the (re)production of urban inequalities. However, climate urbanism emerges with renewed virulence, creating new and unseen forms of climate gentrification and displacement.

Since climate action requires the involvement of various actors with different interests, it implies compromises that prevent a complete overhaul of existing development pathways (Chu et al. 2016). A key feature of climate urbanism is the re-direction of financial investments into technological and low-carbon infrastructural fixes for climate mitigation and adaptation (Long and Rice 2019; Bigger and Millington 2019; Christophers 2018). Such investments are spatially selective, they mobilize climate mitigation and adaptation as normative goals that eventually ensure the survival of neoliberal regimes of capital accumulation and reinforce existing social, racial, economic, and gender-based forms of exclusion. Climate precarity—exclusions from safe zones, exposure to impact, and limited access to low-carbon technologies—is a ubiquitous condition under the shadow of the New Climate Urbanism (see Chap. 3). Climate adaptation planning strategies have been shown to exacerbate dynamics of gentrification and displacement (Anguelovski et al. 2016; Chu et al. 2017; Shi et al. 2016) often resulting in the uneven exposure of different groups of urban dwellers to climate risks (Anguelovski et al. 2019). Actions for the protection of the core economic functions of cities have secured ecological enclaves for the rich (Caprotti 2014; Goh 2019b; Graham and Marvin 2001; Hodson and Marvin 2009, 2017; Sanzana Calvet 2016). Academics and activists have warned against the translation of these actions into a climate apartheid, creating new divides between the climate privileged and the climate precarious (Bond 2016; Rice et al. 2020, Chap. 3). Inhabitants of cities facing large infrastructure deficits are likely to bear the burden of climate change. Climate change impacts urban areas that have contributed little to global greenhouse emissions, particularly those located in the global South (Silver 2018; Mulugetta and Castán Broto 2018). This became visible for instance when cyclone Idai hit the coast of Mozambique, devastating the coastal cities of Beira and Pemba. As discussed here, existing accounts tend to show that climate action in cities follows a neo-colonial logic of enclosure and capital accumulation, marginalizing and dispossessing subaltern groups. But is there anything else to climate urbanism? Can urban areas open up spaces for hope and experimentation? Undoubtedly, many iterations of climate urbanism reinforce the forms of oppression and injustices brought about by the capitalist city. But alongside those, urban communities also engage in multiple forms of experimentation and reimagination of urban futures (Bulkeley et al. 2014). So, while we need a forceful critique of forms of climate urbanism that reproduce the capitalist city, we also need a lens of analysis that enables us to see beyond the critique, and to look into the possibilities of emancipatory and just climate action in cities. Such perspective examines what is overlooked, ignored, and othered. In the next section, we find inspiration in postcolonial theory and decolonial thinking to develop a perspective on climate urbanism that hopes to be both critical and emancipatory.

2.3 Postcolonial Thinking as a Way of Seeing, Decoloniality as a Research Praxis

Postcolonial theory allows us to analyse climate urbanism as the result of past injustice and current dynamics of oppression. Decolonial scholarship recognizes this legacy and invites us to reflect upon how we, as individuals, contribute to the reproduction of colonial thinking and doing. Drawing on both strands of thought, we offer four principles that, we believe, can help thinking through alternatives to current theorizations of climate urbanism and can inform research praxis oriented towards transformative collective action.

The first principle consists in recognizing the links between the global history of imperialism and the deep inequalities of our current world order. There is a common narrative of colonization, as it began in the sixteenth century with the conquest of Latin America and expanded through the occupation of territory, extraction of resources, and enslavement of peoples of the Americas, Africa, Asia, and Oceania by European powers. This narrative assumes that direct colonial control ended through political independence gained by nation-states throughout the twentieth century. However, colonialism extends into the now through the imposition of political institutions, privatized economic relations, and socio-cultural ideals. Mignolo (2017) describes this as a colonial matrix of power—a superstructure that evolved over 500 years into a system and a logic that orders our lives today. Colonization requires overt violence (Fanon 1963), but it is also enacted in insidious ways, operating through the insertion of images and aspirations into the minds of both colonizer and colonized (Fanon 2008). The decolonial project recognizes that we live in a postcolonial world to develop *praxis* for undoing these relations and imaginaries. In the words of Quijano (2007: 168–169), there is "a colonization of the imagination of the dominated; that [...] acts in the interior of that imagination." This happens, most notably, through the definition of modernity and civilization according to Western standards (Maldonado-Torres 2016; Mignolo and Walsh 2018; Vázquez 2012).

Western notions of modernity continue to shape contemporary discourses, such as those of development (Escobar 1995) and sustainability (Banerjee 2003; Ziai 2015). This has implications for climate urbanism, as current approaches to climate mitigation and adaptation are, for a large part, techno-centric, private-led interventions embedded in the modernist project that sees science and technology as solutions to protect us from nature. Developing a minor theorization of climate urbanism requires the active deconstruction of such hegemonic discourses and practices that "negate, disavow, distort and deny knowledges, subjectivities, world senses, and life visions" (Mignolo and Walsh 2018: 4).

The second principle consists in overcoming dichotomies that permeate practices of othering. The Western philosophical tradition tends to construct the world as hierarchical opposites (man/woman, human/nature, reason/emotion) (Merchant 1980; Plumwood 2004). Relationships of mastery and domination are constructed through this way of thinking, together with internalized notions of superiority and difference. Quijano (2007) shows how colonization institutionalized a typology of social discrimination based on race and ethnicity. This explains how colonization normalized the idea of sub-humanity, for those groups whose expendability was socially accepted. Today, these categories produce narratives that define a majority of the world population as inferior based on ethnicity and skin colour, as well as gender, religion, sexual preference, or nationality (Mignolo 2017: 43). In the language of de Sousa Santos (2015), everyone outside the global abyssal lines of exclusion is an object of dispossession, displacement, and deprivation. This normalization extends into the minds of those who are othered, so that "[l]iving in the zone of sub-humanity means ... that it is normal for everything and everyone, including oneself, to question one's humanity" (Maldonado-Torres 2016: 13). Recognizing practices of othering is a crucial project of postcolonial theory and of decoloniality. Breaking down these dichotomies is a strategy to open up to other categories of thinking and being (Mignolo 2017). This is a struggle that links to feminist efforts to deconstruct gender dichotomies (Butler 2002) and the artificial othering of nature (Mies and Shiva 1993; Plumwood 2004). In relation to climate urbanism, challenging these dichotomies is a strategy to bring into focus the perspectives of those who are most vulnerable to the impacts of climate change, yet whose lives are made more precarious by climate change action (see Chap. 3).

The third principle consists in challenging the existence of a universal form of knowledge and relates to the universalization of Western knowing, which disguises the particularities and subjectivities embedded in this worldview (Quijano 2007). Western cosmology is one among others, which is "different only in the differential of power" (Mignolo 2014: 24). The idea of a universal knowledge relies on the portrayal of non-Western civilizations as backward, ignorant, and irrational (Said 1995). Colonization led to the past and current destruction of multiple forms of knowledge production, beliefs, and expression, which were replaced by those of the West (Quijano 2007). The decolonial project embraces what de Sousa Santos (2015) describes as ecologies of knowledge: a plurality of ways of knowing, valuing, and experiencing the world, especially those expressed by the currently voiceless. Operating within ecologies of knowledge demands sensitivity towards questions such as how multiple and different knowledges can co-exist, communicate, and relate to each other (Walsh 2012). One way to approach this co-existence is through the concept of *vincularidad*, which describes the interconnections and interdependence among all dimensions of life (Mignolo and Walsh 2018). In practice, this may consist of, for example, strategies to actively construct spaces to recall forgotten histories, (re-)construct social imaginaries, revive ancestral knowledge, and build collective identities (Walsh 2012). This kind of *vincularidad* is crucial for climate urbanism research, as nurturing and drawing on multiple forms of knowledge can help both understand and act in a climate-changed urban world (see Chaps. 12, 13 and 14).

The fourth principle posits that while decoloniality is a collective project, it can only ever start with the self. As Fanon (2008) stated, racism and discrimination are socially produced, lodged within the psychology of the individual, and reproduced through everyday social interactions. Reflecting on our self-perception and social interactions must, therefore, always be the starting point for decolonization. In the words of Mignolo and Walsh (2018: 38):

> [i]f you see yourself as a leader who will help others to decolonize, you would be taking the first wrong step … Decolonization is a communal and collective work grounded in the self-awareness of decolonial subjects.

This starting point has several implications for our work as researchers, particularly when it comes to examining intersecting forms of injustices, appropriation, and exploitation causing and resulting from the climate crisis. It challenges the dichotomy between theory and praxis, where decoloniality

> is a way, option, standpoint, analytic, project, practice, and praxis … decolonial thinking is always decolonial doing, and doing decolonially is tantamount to thinking decolonially. (Mignolo and Walsh 2018: 4, 38)

As has been argued by feminist scholars, it is necessary to engage with multiple situated experiences to generate knowledge, and also to appreciate and make visible overlooked worldviews (e.g. Haraway 1988). This principle also relates to how we conduct education and research, for example, how a predominant focus on publications and rankings perpetuate elitism and exclusion (Maldonado-Torres 2016). A minor perspective on climate urbanism also demands attention to decolonizing methodologies (Smith 2013) in what we study, including how we formulate questions, who we work with, how we connect our work with people and problems on the ground, and how we present our work in written and non-written form. We thus propose rethinking climate urbanism from a 'minor' perspective, focusing specifically on marginalized forms of knowledge and mundane attempts to adapt to and cope with a changing climate that we see emerging in various cities of the global South and North. Our contention here is that a minor perspective on climate urbanism will help us to recognize how climate action in cities is shaped by colonial and neoliberal forms of domination, but it can also help us understand how alternatives to these models emerge, and can give us the tools to take part in those alternatives.

2.4 Decolonizing Climate Urbanism

The dominant focus on neoliberal and neo-colonial processes of accumulation through urban climate policy have paradoxically contributed to closing the debate on what constitutes just and progressive forms of climate urbanism. We still lack an understanding of how situated forms of climate urbanism emerge embedded in the lives of urban citizens, in changing cultures, and in ways of living the city which clash against (or at least question) neoliberal models of climate urbanism (Robin and Castán Broto forthcoming). Climate-proof spatial plans and infrastructural fixes happen alongside other more mundane, and more just, forms of climate action. Such initiatives can be found, for example, in environmental justice movements (e.g. Bullard 1999, 2000; Pulido 1996) or in social movements motivated by climate justice (Schlosberg and Collins 2014). These social movements offer radically new ways of looking at cities' economic, racial, and socio-ecological relations in a climate-changed world. What is progressive, or even just, in terms of climate action cannot be defined in advance, but certainly relates to the operation and results emerging from multiple forms of climate politics. Climate urbanism research can offer a diagnosis of current issues alongside opening up spaces of possibility for alternatives as long as it tunes into the diversity of urban responses to climate change in mundane contexts of action. To be sure, no climate strategy is inherently progressive or without contradictions. However, efforts to deliver inclusive and collective urban futures on the ground are often rooted in attempts to challenge the hegemonic discourses on technocratic or finance-led approaches to climate-resilient cities. A comprehensive climate urbanism research agenda should engage with the multiple locations of climate action, often perceived as marginal. In other words, building a minor theory will require us to look into those spaces that are obscured behind the shine of fashionable and widely promoted climate action. Table 2.1 offers a starting point to think through a

Table 2.1 Principles for a minor perspective on climate urbanism

Principle for a postcolonial way of seeing	Postcolonial diagnosis	Decolonising practice
Historical link between the global history of imperialism and the deep inequalities of the current world order	Climate urbanism emerges within the remnants of a colonial matrix of power	Identify and make visible how the legacy of colonial projects nurtures climate precarity and vulnerabilities in cities. Uncover explicit and implicit forms of colonial violence and how they are performed within distinct versions of climate urbanism in different places.
Overcoming dichotomies that permeate practices of othering	Climate urbanism is implemented and reinforced through practices which divide the city in a deserving and an undeserving one, one which is divided in different territories of safety and investment. Climate apartheid is grounded in the ideas of securitization and division that perpetuate elitism and exclusion.	Recognise the multiple opportunities for building safety, create people-oriented responses to climate change that also works with the non-human. The process starts with an inclusive, collective vision of the urban capable to mobilise non-extractive financial resources such as participatory budgeting and other bottom-up initiatives for climate action.

(*continued*)

Table 2.1 (continued)

Principle for a postcolonial way of seeing	Postcolonial diagnosis	Decolonising practice
Challenging the existence of a universal form of knowledge	Climate urbanism follows on from a climate of distrust and deliberate misrepresentation of climate change through denialism, putting science in a defensive position, and privileging techno-centric 'fixes,' which negate minor forms of existence, knowledge, and visions. Knowledge about climate change responses and their effectiveness needs to be open to deliberation and collaboration that integrate multiple forms of knowledge.	The recognition of multiple knowledges goes hand in hand with a celebration of 'vincularidad' in climate urbanism which has to be negotiated through the active inclusion of those whose knowledge is denied. It demands an intersectional diagnosis of climate vulnerabilities and of different groups' participation in climate action. The approach recognises alternative modes of organising urban society under climate change, many of those rooted in subaltern experiences.
The collective project of decolonisation starts with the self	Reflection on individual praxis, emotional engagement and ethics of care should be the starting point for collective action and research.	A collective, emancipatory, climate urbanism project requires most of all a sense of collective hope mobilised to work together to a future, shared city - this includes sharing the urban environment with the non-human.

minor perspective on climate urbanism, one grounded in postcolonial theory and rooted in the aims of decolonial praxis, using the four principles outlined in the previous section as a critical frame for analysing climate action and decolonising research practices.

2.5 Conclusion

Our objective in this chapter is to invite other scholars, activists and researchers to partake in efforts to build a minor perspective on climate urbanism: a research agenda that recognizes how neo-colonial, patriarchal and capitalist relations of power shape the possibilities for climate action in cities, while searching for multiple alternatives that help us reimagine a collective future. The question that remains is whether orienting climate urbanism research towards the development of a minor theory and a commitment to decolonial aims will be sufficient to dismantle neo-colonial and capitalist forms of oppression, and to ensure the safety of the most vulnerable to climate change impacts. This is no small task, and undoubtedly beyond what academic scholarship alone is and will ever be able to achieve. However, as we contend throughout this chapter, a first step would be to recognize that climate urbanism emerges from multiple sites, led by a diverse array of actors, and delivering ambiguous results. Future research should thus reorient itself towards the integration of the multiple and seemingly minor activities taking place in cities. However, this will not be sufficient if not linked to a commitment to undoing various forms of oppression, through nurturing collaborations within and beyond academia, and through the extension of academic networks beyond those that produce hegemonic knowledge.

References

Anguelovski, I., Shi, L., Chu, E., Gallagher, D., Goh, K., Lamb, Z., et al. (2016). Equity impacts of urban land use planning for climate adaptation: Critical perspectives from the global north and south. *Journal of Planning Education and Research, 36*(3), 333–348.

Anguelovski, I., Connolly, J.J., Pearsall, H., Shokry, G., Checker, M., Maantay, J., Gould, K., Lewis, T., Maroko, A. & Roberts, J.T. (2019). Opinion: Why green "climate gentrification" threatens poor and vulnerable populations. *Proceedings of the National Academy of Sciences, 116*(52), 26139–26143.

Banerjee, S. B. (2003). Who sustains whose development? Sustainable development and the reinvention of nature. *Organization studies, 24*(1), 143–180.

Bigger, P., & Millington, N. (2019). Getting soaked? Climate crisis, adaptation finance, and racialized austerity. *Environment and Planning E: Nature and Space*: 251484861987653.

Bond, P. (2016). Who wins from "Climate Apartheid?": African climate justice narratives about the Paris COP21. *New Politics, 15*(4), 83.

Bulkeley, H. A., Broto, V. C., & Edwards, G. A. (2014). An urban politics of climate change: experimentation and the governing of socio-technical transitions. Routledge.

Bullard, R. D. (1999). Dismantling environmental racism in the USA. *Local Environment, 4*, 5–19.

Bullard, R. D. (2000). *Dumping in Dixie: Race, class, and environmental quality* (3rd ed.). Boulder, Colorado: Westview Press.

Butler, J. (2002). *Gender trouble*. Abingdon: Routledge.

Caprotti, F. (2014). Eco-Urbanism and the Eco-City, or, denying the right to the city?: Eco-Urbanism and the Eco-City. *Antipode, 46*(5), 1285–1303.

Christophers, B. (2018). Risk capital: Urban political ecology and entanglements of financial and environmental risk in Washington, DC. *Environment and Planning E: Nature and Space, 1*(1–2), 144–164.

Chu, E., Anguelovski, I., & Carmin, J. (2016). Inclusive approaches to urban climate adaptation planning and implementation in the Global South. *Climate Policy, 16*(3), 372–392.

Chu, E., Anguelovski, I., & Roberts, D. (2017). Climate adaptation as strategic urbanism: Assessing opportunities and uncertainties for equity and inclusive development in cities. *Cities, 60*, 378–387.

de Sousa Santos, B. (2015). *Epistemologies of the South: Justice against epistemicide*. London: Routledge.

Escobar, A. (1995). *Encountering development. The making and unmaking of the third world*. Princeton, New Jersey: Princeton University Press.

Fanon, F. (1963). *The wretched of the earth*. New York: Grove Press.

Fanon, F. (2008). *Black skin, white masks*. New York: Grove press.

Goh, K. (2019a). Flows in formation: The global-urban networks of climate change adaptation. Urban Studies, 004209801880730.

Goh, K. (2019b). Urban waterscapes: The hydro-politics of flooding in a sinking city. *International Journal of Urban and Regional Research, 43*(2), 250–272.
Graham, S., & Marvin, S. (2001). *Splintering urbanism: Networked infrastructures, technological mobilities and the urban condition*. London; New York: Routledge.
Haraway, D. (1988). Situated knowledges: The science question in feminism and the privilege of partial perspective. *Feminist studies, 14*(3), 575–599.
Hodson, M., & Marvin, S. (2009). "Urban ecological security": A new urban paradigm? *International Journal of Urban and Regional Research, 33*(1), 193–215.
Hodson, M., & Marvin, S. (2017). Intensifying or transforming sustainable cities? Fragmented logics of urban environmentalism. *Local Environment, 22*, 8–22.
Katz, C. (1996). Towards minor theory. *Environment and Planning D: Society and Space, 14*(4), 487–499.
Long, J., & Rice, J. L. (2019). From sustainable urbanism to climate urbanism. *Urban Studies, 56*(5), 992–1008.
Luque-Ayala, A., Marvin, S., & Bulkeley, H. (Eds.). (2018). *Rethinking urban transitions: Politics in the low carbon city*. Abingdon, Oxon; New York: Routledge, Taylor and Francis Group.
Maldonado-Torres, N. (2016). Outline of ten theses on coloniality and decoloniality. Retrieved from Foundation Frantz Fanon: http://frantzfanonfoundation-fondationfrantzfanon.com/article2360.
Merchant, C. (1980). *The Death of Nature: Women, Ecology and the Scientific Revolution*. New York: Harper Collins.
Mies, M., & Shiva, V. (1993). *Ecofeminism*. London: Zed Books.
Mignolo, W. D. (2014). Further thoughts on De(Coloniality). In S. Broeck & C. Junker (Eds.), *Postcoloniality-decoloniality-black critique: Joints and fissures* (pp. 21–52). Frankfurt: Campus Verlag.
Mignolo, W. D. (2017). Coloniality is far from over, and so must be decoloniality. *Afterall: A Journal of Art, Context and Enquiry, 43*(1), 38–45.
Mignolo, W. D., & Walsh, C. E. (2018). *Introduction, on decoloniality: Concepts, analytics, praxis* (pp. 1–12). Durham: Duke University Press.
Mulugetta, Y., & Broto, V. C. (2018). Harnessing deep mitigation opportunities of urbanisation patterns in LDCs. *Current opinion in environmental sustainability, 30*, 82–88.

Plumwood, V. (2004). Gender, eco-feminism and the environment. In R. White (Ed.), *Controversies in environmental sociology*. Cambridge: The University of Cambridge.

Pulido, L. (1996). *Environmentalism and economic justice: Two Chicano struggles in the Southwest, Society, environment, and place*. Tucson: University of Arizona Press.

Quijano, A. (2007). Coloniality and modernity/rationality. *Cultural studies, 21*(2–3), 168–178.

Rapoport, E. (2014). Utopian visions and real estate dreams: The eco-city past, present and future: Utopian visions and real estate dreams. *Geography Compass, 8*(2), 137–149.

Rashidi, K., Stadelmann, M., & Patt, A. (2019). Creditworthiness and climate: Identifying a hidden financial co-benefit of municipal climate adaptation and mitigation policies. *Energy Research and Social Science, 48*, 131–138.

Rice, J. L., Cohen, D. A., Long, J., & Jurjevich, J. R. (2020). Contradictions of the climate-friendly city: New perspectives on eco-gentrification and housing justice. *International Journal of Urban and Regional Research, 44*(1), 145–165.

Robin, E. & Castán Broto, V. (forthcoming). Towards a postcolonial perspective on climate urbanism. *International Journal of Urban and Regional Research.*

Said, E. W. (1995). *Orientalism: Western conceptions of the orient, with a new afterword.* London: Penguin Books.

Sanzana Calvet, Martin. (2016). The greening of neoliberal urbanism in Santiago de Chile: Urbanisation by Green Enclaves and the Production of a New Socio-Nature in Chicureo, 443.

Schlosberg, D., & Collins, L. B. (2014). From environmental to climate justice: Climate change and the discourse of environmental justice. *Wiley Interdisciplinary Reviews: Climate Change, 5*, 359–374.

Shi, L., et al. (2016). Roadmap towards justice in urban climate adaptation research. *Nature Climate Change, 6*(2), 131–137.

Silver, J. (2018). Suffocating cities: Urban political ecology and climate change as social-ecological violence. Urban political ecology in the anthropo-obscene: Interruptions and possibilities, 129–146.

Smith, L. T. (2013). *Decolonizing methodologies: Research and indigenous peoples.* London: Zed Books Ltd.

Vázquez, R. (2012). Towards a decolonial critique of modernity. Buen Vivir, relationality and the task of listening. *Capital, Poverty, Development, Denktraditionen im Dialog: Studien zur Befreiung und interkulturalität, 33*, 241–252.

Walsh, C. (2012). "Other" knowledges, "Other" critiques: Reflections on the politics and practices of philosophy and decoloniality in the "Other" America. *Transmodernity, 1*(3), 11–27.

Ziai, A. (2015). *Development discourse and global history from colonialism to the sustainable development goals.* London: Routledge.

3

Climate Urbanism and the Implications for Climate Apartheid

Joshua Long, Jennifer L. Rice, and Anthony Levenda

3.1 Introduction

Within the first two decades of the twenty-first century, climate action steadily transitioned from a speculative conversation piece among urban policymakers to a nearly ubiquitous planning priority among leaders in the world's largest cities. As this transition has unfolded, it has become increasingly evident that implementing climate policies requires an unprecedented mobilization of financial and material resources to fund new programmatic, technological, and infrastructural solutions to the

J. Long (✉)
Southwestern University, Georgetown, TX, USA
e-mail: jlong@southwestern.edu

J. L. Rice
University of Georgia, Athens, GA, USA
e-mail: jlrice@uga.edu

A. Levenda
University of Oklahoma, Norman, OK, USA
e-mail: anthonylevenda@ou.edu

V. Castán Broto et al. (eds.), *Climate Urbanism*,
https://doi.org/10.1007/978-3-030-53386-1_3

climate crisis. Priority projects associated with these solutions include examples such as defensive sea walls, new digital infrastructures and platforms, exclusive self-sustaining eco-districts, and the physical reinforcement of institutions, transportation networks, utility infrastructures, and industries deemed necessary for the reproduction of neoliberal capitalism. These actions represent a dominant mode of climate urbanism. Situated within a narrative that emphasizes an urgent reaction to the threats posed by climate change, the rise of climate urbanism presents a complex and contested landscape for policymakers, planners, and citizens. A central question of this era will be whether climate urbanism materializes as democratic and transformative or as exclusive and segregated. With that question in mind, this chapter introduces a theoretical framework that more deeply situates the political and economic context of climate urbanism. More importantly, it presents a critical argument that suggests any equitable, just, and transformative action must first face the socio-structural foundations of the climate crisis.

In what follows, we discuss the transition from the era of sustainable urbanism to the rise of climate urbanism, define and discuss the dominant features of climate urbanism, and finally discuss the importance of broadening this framework to develop a critical lens that can be used to dismantle the root causes of the climate crisis. Throughout much of this chapter, we refer largely to a dominant narrative of climate urbanism—one that is materializing within the context of neoliberalism, and thus far, remains immersed in western, modernist approaches to development. As Robin and Castán Broto (forthcoming) have noted, however, alternative narratives and perspectives are beginning to emerge and have the capacity to move toward a mode of climate urbanism that prioritizes justice.

3.2 From Sustainable Urbanism to Climate Urbanism

By the start of the twenty-first century, neoliberal growth strategies and the principles of sustainable urban development had converged to produce a set of policy initiatives that came to define an era of urban greening in North America and much of Europe. This was the period of sustainable urbanism, one in which a "sustainability fix" (While et al. 2004) emerged to balance economic and environmental concerns according to the interests of policymakers and business leaders. The sustainability fix was accompanied by a narrative of green growth that prioritized "win-win" programs and policies that sought to reconcile the tensions between economic growth and sustainability. These included ecological modernization, density-oriented development, transit-oriented development, and smart city solutions—initiatives that increased city marketability, appealed to capital investment, and provided a desirable set of amenities for creative and tech workers who were increasingly valorized as the greatest innovators and economic contributors of the labor force (While et al. 2004; Luque-Ayala and Marvin 2015; Rice et al. 2020). The economic and environmental wins of sustainable urbanism were expected to have valuable implications for social sustainability, but the greening of neighborhoods, the purposeful attraction of tech-oriented, creative industries and workers, and technological modernization largely led to housing (un)affordability crisis, economic polarization, displacement, and environmental gentrification (Quastel 2009; Checker 2011; Long 2016). This continued through the financial crisis of the early twenty-first century, an event that spread internationally and heightened problems of displacement and inequality in major North American and European cities. As the financial crisis spread globally, this pattern would reproduce itself in cities in Latin America, Australia, and East Asia. As Rolnik (2013) notes, the post-crisis era saw neoliberal strategies being employed in different ways globally through the continued financialization of home-ownership, the "unlocking" of urban land values, and the privatization of public spaces alongside new modes of austerity governance. This is particularly important for understanding the rise of climate urbanism. In the post-financial crisis era, cities have remained crucial "strategic targets and proving grounds" for "policy experiments, institutional innovations, and political projects" (Theodore et al. 2011: 24) at the same time that climate change has become a priority for planners and policymakers.

While nascent discussions of climate policy were relatively weak at the outset of the twenty-first-century, by the 2010s, climate action had become a widespread discussion among policymakers and planners in almost every major city in the world (Watts 2017). As a result, numerous cities were adopting plans to reduce emissions while also proclaiming themselves as champions in the fight against climate change (Rice 2014; Watts 2017). As city leaders increased emissions reduction efforts, they likewise realized the importance of climate adaptation by protecting infrastructures, institutions, and services from the threats posed by climate hazards. This concern was shared at multiple scales. Within a relatively short period of time, interurban networks, development agencies, and global financial institutions began to synthesize a narrative of climate action that emphasized several key components including the prioritization of climate resilient infrastructure, the promotion of technological innovation, the provision of networks for the exchange of technological expertise and support, and the advancement of new financial and logistical pathways for investment in climate-oriented programs (Long and Rice 2019). It was a narrative that brought together parallel discussions of sustainability, climate mitigation, and adaptation under a malleable framework of "climate resilience" (see Chap. 8). In doing so, it not only re-appropriated the climate crisis to fit within a neoliberal framework of development in a "new urban climate economy" (Floater et al. 2014; GCEC 2016), it also spoke directly to public angst over climate change and introduced an existential sense of urgency that trumped competing concerns of inequitable development and environmental justice. Further, this narrative opened up the potential for creative experimentation in exclusive housing projects focused on resiliency (discussed later in this chapter), expanding the market for retreat opportunities for the super-rich, whose transnational search for insulated, climate-hardy, and "heavily secured environments" has disrupted housing markets on a global scale (Paris 2013). In the next section, we define climate urbanism while also briefly explaining how it fits within some of the major trends in neoliberal urban governance. We then explain in detail the potential consequences of its accompanying political narrative of defensive and reactive climate action.

3.3 Defining and Deconstructing Climate Urbanism

Climate urbanism can most simply be defined as a twenty-first-century policy orientation that (1) promotes cities as the most viable and appropriate sites of climate action and (2) prioritizes efforts to protect the infrastructures and institutions of urban economies from the hazards associated with climate change (Long and Rice 2019). This foundational definition, however, must be considered within the context of its key tenets, and the political narrative used to justify its adoption. We recognize—and are hopeful—that the dominant discourse of climate urbanism may evolve to become more inclusive and transformative. Portraying the dominant rhetoric of climate urbanism as a uniform set of actions overlooks the alternatives emerging outside of circuits of power in the global North (see Chap. 2). So, while we offer a simplified set of features here, we recognize that these have materialized specifically within a unique context that itself requires interrogation (*ibid.* 994):

(1) Climate urbanism hinges upon the presumption that cities and local municipalities are the most capable and credible sites for addressing climate change and implementing climate governance through carbon control;
(2) Climate urbanism prioritizes technological and infrastructural projects as the most important solutions to the climate crisis;
(3) Climate urbanism promotes new institutional frameworks and global investments to finance these otherwise cost-prohibitive solutions.

To begin, focusing on the urban as the primary scale of intervention is, in many ways, a practical choice. Cities contain the majority of the global population, they are responsible for as much as 80% of global gross domestic product, and many of them are located in areas that are particularly susceptible to climate hazards like rising sea levels, increased storm intensity, flooding, and drought (Dobbs et al. 2011; Estrada et al. 2017). This situation makes cities logical choices for the protection of vulnerable populations, institutions, and infrastructures. Further, while there has been relative inaction on climate policy at the supranational level, many cities have moved more quickly to address climate change (Watts 2017). As such, cities emerged as a site where the political will to take action could be mobilized over various activities subject to local governance (e.g. transit, building codes, infrastructure). At the same time, this focus on the urban scale dovetails neatly with neoliberal promotion of interurban competition, tightly controlled capital investments, localized public-private partnerships, and landscapes of technological and market experimentation. Ultimately, there is widespread recognition among many policymakers that cities are the preferred scale for climate action and capital generation in the so-called "new climate economy." Simply put, the concentration of wealth, influence, and population ensures that protection of the infrastructures and institutions of cities is the most pragmatic way to hedge against climate hazards, and investment in such projects is also viewed as a way to generate profit during the climate crisis.

This is potentially problematic for multiple reasons. First, there is little evidence to suggest that—in a market-driven neoliberal world that prioritizes projects offering the highest return on investment—the protection of infrastructures and institutions will materialize in an equitable way within cities. In order for these projects to produce transformative and just results, the decision-making process must be democratized and must prioritize social justice over the concerns of elites and the politically influential (Rice 2016). Second, the focus on infrastructural and technological solutions introduces questions about who controls and profits from those capabilities. Currently, the expertise and immense financial resources needed to implement these projects can only be found among the largest technological conglomerates, financial institutions, and government agencies of the wealthiest countries—a feature that supports the reproduction of a particular mode of corporate capitalism concentrating data control and network management in the hands of a few (Fields, Bissel and Macrorie 2020). This situation suggests that disparities resulting from climate urbanism exist not only within cities, but comparatively between wealthier and poorer cities. Lastly, the immense resources needed to finance these projects (estimated at approximately USD 6.3 trillion per year by the World Bank) requires credible financial networks and pathways for mobilizing and channeling this money into projects. These are emerging, and are being orchestrated by organizations such as the

World Bank, IMF, OECD Resilient Cities, United Nations Green Climate Fund, C40 Cities Climate Leadership Group, and numerous others. While some are new, many of these organizations have a troublesome historical record when it comes to managing development initiatives in an equitable, ethical, and sustainable manner. For instance, organizations such as the World Bank, OECD, and IMF have been heavily criticized for exacerbating international debt, perpetuating colonial legacies of power and political control, and promoting initiatives that have led to significant environmental degradation (Demissie 2007; Peet 2009). A unified narrative of market-driven and growth-oriented climate action has become formalized within the institutional documents and promotional messaging of these organizations. That narrative is urgent, reactive, and predicated on a moral imperative that emphasizes swift action to ensure human survival. It oversimplifies the complicated environmental and social challenges that accompany climate action. Rather than focusing on the complexity of ecological issues such as biodiversity loss, pollution, agricultural sustainability, and others, the logic of climate urbanism prioritizes the control of greenhouse gases and focuses on a defensive mode of climate resilience. Instead of addressing the social problems highlighted in the era of sustainable urbanism (i.e. human rights violations, income inequality, segregation, marginalization), the narrative of urgency recasts these issues under a new moral imperative that trivializes their salience against the threat of planetary collapse, human survival, and security.

To be clear: we agree that climate action is urgently needed, infrastructural and technological solutions are vital to municipal climate action plans, and there will be a need to mobilize significant financial and material resources to mitigate and adapt to a changing climate. However, a truly transformative, ethical, and democratic approach to climate urbanism is needed to ensure that climate-oriented policies and projects benefit the global population in a just and equitable way. The current trajectory of climate urbanism offers little evidence that this is occurring. Thus far, the prioritization of infrastructure projects has been focused less on projects that provide humanitarian relief or assistance, and more on projects that reproduce and/or protect capitalist economies, or defensive measures done in the name of security and surveillance. It is imperative that climate urbanism does not lead to a form of climate fascism that justifies the rights of some nations and peoples over others in the name of ecological safety (Huq and Mochida 2018). Currently, climate disasters are displacing millions of Indigenous and poor populations of developing countries at the same time that anti-immigrant rhetoric and violence is escalating in wealthy and mostly majority-white nations—a phenomenon that does not bode well at all for human rights and social justice. In many cities of the Global North, the era of climate urbanism has begun with the announcement of new and exclusive techno-utopian enclaves of climate resilience at the same time that the most vulnerable populations are being dismissed as unfortunate casualties of the climate crisis.

We suggest that there are specific reasons for this, and that in order to achieve any measure of transformative climate justice, the root causes of the climate crisis must be addressed. As Silver (2018: 7–15) notes, the urban spatial separation that has persisted since the colonial era is being reproduced through the climate crisis:

> The violence of climate change is therefore every bit as explicit and powerful in its spatial demarcation as colonialism, for its origins lie in the very same processes of underdevelopment and racial capitalism.

In the following section, we discuss the importance of widening the lens of climate urbanism in order to focus on the underlying historical structures of dehumanization and violence that have led to the climate crisis.

3.4 Against Climate Apartheid and Toward a Transformative Climate Urbanism

Scholars and activists working on climate policy have seen little meaningful progress toward addressing the way climate change has disproportionately impacted vulnerable populations. Further, we have witnessed the sharp imbalance in how urban governance has either intervened on their behalf or planned for their future. We suggest that this is because much of the work to solve the climate crisis has been focused on policy measures that address the material causes of global warming (i.e. carbon reduction, energy transitions, trading emission schemes, carbon capture). However, even though such policy measures are necessary for climate intervention, they do not address the root causes of the climate crisis, which are inseparable from the structures that continue to sustain racism, dehumanization, and violence against vulnerable and marginalized populations.

In this section we broaden the critical lens used to interrogate climate urbanism in hopes of naming, deconstructing, and dismantling a contemporary iteration of global apartheid exacerbated by the climate crisis.

We use the term "climate apartheid" to describe this, and we should mention that we are not the first to employ this term. Indeed, Bond (2016) uses "climate apartheid" to address a political ecology of uneven climate impacts in Africa, Dawson (2017) uses it to draw awareness to the vulnerability of climate displaced migrants and refugees, and Tuana (2019) uses the term "Climate Change Apartheid" to point to the disproportionate impacts of climate change deeply rooted in legacies of racism and exploitation. We follow these scholars and others as we draw from engagements with feminist, race and ethnicity studies, and human geography to deconstruct the historical legacies and contemporary structures of settler colonialism, racial capitalism, and hetero-patriarchy. We argue that these three interlocking systems have exacerbated the climate crisis by normalizing the dehumanization, exploitation, and dispossession of historically oppressed communities while simultaneously promoting landscapes, infrastructures, and policies of security and isolation from climate hazards for privileged populations. We refer to this system of privilege and precarity in our definition of climate apartheid: a global system of discrimination, segregation, and violence built upon constructions of race, class, and gender that uses climate change—and efforts to address the climate crisis—to justify and reproduce itself.

It is important to recognize that current configurations of climate apartheid are tied to historical and ongoing processes. The first of these is settler colonialism, as the act of colonization is not an event of the past, but continues today through the ongoing occupation of Indigenous lands by settler communities. Violence against Indigenous communities, and the land dispossession that accompanied that violence, continues as a logic of elimination that normalizes the death of non-white populations for the benefit of settler communities (Dotson 2018). Exacerbated under climate change, the expendability of some communities who are subjected to climate-related harm and death can be seen as an extension of this logic.

Relatedly, climate apartheid is underpinned by systems of racial capitalism, which derives economic value through the processes of differentiation and racialization (Robinson 2005). While examinations of capitalism explain differences in class, *racial* capitalism helps us understand how and why racial differentiation is both central to and preexisted neoliberal capitalism. Put in the context of climate urbanism, racial capitalism allows us to understand how carbon-intensive economic development is tied to the forms of environmental racism that exposes racialized bodies to premature death.

Finally, we recognize that climate apartheid is related to the persistence of heteropatriarchy, which values a specific set of technocratic and scientific knowledge over others, and validates continued domination of the environment and climate system (see Chap. 2). Continued supremacy of a cis, heterosexual, masculine socialized group reinforces a specific anthropocentric hierarchy of planetary dominance and degradation. This overshadows an ecocentric perspective espoused by many non-western peoples, while simultaneously creating a hierarchy within human society. Such a dominant perspective undervalues the experiences and needs of marginalized groups, while further exposing female, poor, queer, and Indigenous communities to more of the harms of climate change (Buckingham and Le Masson 2017; Moosa and Tuana 2014).

Naming and deconstructing these root causes of climate apartheid are key discursive practices needed to inform activism and policies. It is also important to remember that discursive engagements with structures help to increase awareness about everyday practices of segregation, displacement, and violence. These practices are ongoing and occurring in multiple ways. First, there is a wealth of evidence that reveals climate hazards disproportionately impact poor, marginalized, and vulnerable populations as they are more susceptible to these hazards (Sultana 2014; Thomas et al. 2019). Second, the geopolitics of access to resources and ecosystems represents an observable manifestation of climate apartheid, one that showcases the vulnerability of women and Indigenous groups while simultaneously exhibiting the extraterritorial power of wealthy nation-states. As contemporary land grabs in Sub-Saharan Africa, the Arctic, Southeast Asia, and Latin America reveal, both a symbolic and literal erasure of populations and land tenure is occurring through the reproduction of a frontier mentality. Third, and closely related to climate urbanism, there is an array of new infrastructure projects that are climate-conscious, tech-centric, defensive, and often segregating in their social implications. These include projects such as China's South-North Water Transfer project (Webber et al. 2017), South Korea's Saemangeum Sea Wall (Hahnhee 2010), New York City's Lower Manhattan Coastal Resiliency Project (DuPuis and Greenberg 2019), and numerous others. In the case of many projects in global South cities, the potential for displacement and segregation is overshadowed by a public narrative of resilience, progress, and technological advancement. These include examples such as Jakarta's Great Garuda Sea Wall (Goh 2019) and many planned African cities such as Nairobi's Konza Techno City, Accra's Hope City, or Cape Town's WesCape (Silver 2014). But the examples also include entirely new cities such as Lagos' Eko Atlantic—an eco-island development off the coast of Lagos that is being hailed as a climate-resilient smart city development that will mitigate against storm surges, sea level rise, and coastal erosion (Ajibade 2017). Like some cities already mentioned, this phenomenon also includes tech-centric "fast cities" such as Songdo, Masdar, and Rajarhat that are presented as exclusive, "smart," and eco-friendly urban mega-projects (Datta and Shaban 2016).

Of course, while we discuss existing and proposed developments, it is as important to emphasize the problematic nature of the *narrative* of climate apartheid. As other scholars have suggested, a dual narrative of utopian resilience and climate apocalypse are being used to characterize a "post-political" landscape of segregation and securitization while also redefining notions of citizenship (Swyngedouw 2013; Davoudi 2014; Chaturvedi and Doyle 2015). An eco-utopian narrative of resilience portrays new urban landscapes of technological advancement, efficiency, and climate-defensive infrastructure serving as resilient retreats for the climate privileged. This is juxtaposed against a second narrative that depicts disastrous landscapes of biodiversity loss, competition for dwindling food and material resources, and desperate mass migration of the climate precarious. The simultaneous employment of these narratives facilitates—and attempts to justify—the development of highly segregated "extreme cities" (Dawson 2017) or "elysian sanctuaries … for the new environmental elites" (Brisman et al. 2018: 9). This, of course, has much broader scalar implications for climate migrants/displaced peoples, nation-state borders, national security, and neocolonialism. In addition, citizenship and migration in the era of climate change are highly precarious, introducing questions of mobility, refugee status, and asylum amid increased anti-immigration sentiment and militarization of borders. Despite increasing concerns over "border security" in the developed North, the ability of elites to disregard international borders has become streamlined in many countries, where citizenship and access can be purchased for a high price. This ability to emphasize or disregard borders is a privilege that is also practiced by nation-states, who have begun to argue for a type of climate change era *lebensraum* in the pursuit of resources in poorer countries and "frontier" territories such as Greenland, the Arctic, and Antarctica.

It is important to reiterate why these developments represent a global and climate-induced mode of apartheid. Like South African apartheid and other historical iterations of apartheid, climate apartheid is the culmination of a systematic, historical project of dehumanization, dispossession, violence, and exploitation motivated by the supremacist ideologies of settler colonialism, executed by the instruments of racial capitalism, and normalized through the dominant lens of hetero-patriarchy. The arrival of the climate crisis accelerates this project by portraying an existential and urgent threat, thereby juxtaposing the supremacist case for survival of the chosen against the potential extinction of the human race. Even when rendered visible by economic mechanisms that discriminate, built environments that segregate, and security policies that exclude and criminalize, the separation of the climate privileged and the climate precarious is perversely supported by a narrative of perseverance and survival. This is why we argue that it is imperative to name climate apartheid, deconstruct its structural roots, and present strategies for its abolition. As the current embodiment of an historically self-reinforcing cycle of privilege and precarity, even concrete policies for climate mitigation are susceptible to apartheid unless constructive avenues for intervention are presented. These systems were built upon deep historical relationships that have been, and still are, dependent on a colonial mindset of extraterritorial violence and dispossession. Our contemporary system of racial capitalism—a system that necessitates the dehumanization, criminalization, and exclusion of non-white bodies to justify economic growth for its proponents—is evocative of the violence of previous apartheids and genocides.

Charting these injustices leaves the question: *what is to be done?* Cautionary stories and trajectories about climate apartheid leave us looking for alternative forms of climate urbanism. Inspiration from emerging movements on climate justice rooted in anti-racism, anti-capitalism, Indigenous resurgence, and more, suggest a trend toward radical change. Activists and scholars are taking abolitionist stances more directly in response to the technocratic, inequitable, and post-political machine of urban climate response. Recent work on abolition ecologies (Pulido and De Lara 2018; Heynen 2016) and abolitionist climate justice (Ranganathan and Bratman 2019) foregrounds the necessity of investigating the historical production of climate injustice and oppression. For example, climate response typically considers social dimensions of climate change in terms of vulnerability as a set of predictive demographic categories (e.g. race, income, education level). These measures of vulnerability, however, fail to account for the racist, classist, and sexist underpinnings of the climate crisis that create differentiated vulnerability in the first place. As a result, several scholars show how anti-racist and anti-colonial epistemologies might resist this depoliticization. Jacobs (2019) argues that radical planning and Black feminism can help us center knowledge and action rooted in groups fighting these oppressions to reject normative framings of climate disaster that dismiss community knowledge and experiences. Brand (2018: 18) suggests that there is a space of resistance that "exposes the white supremacist undercurrents of property and capital" and recasts formerly oppressed bodies and spaces as "resilient and beautiful rather than blighted and abandoned." Additionally, unscripted tactical interventions of "guerrilla urbanism" (Hou 2020), and the increasing visibility of Indigenous activism for climate justice

(Norman 2017) are occurring amid heightened awareness of hegemonic neoliberal practices—a phenomenon that is seemingly fueling emergent alternatives. And lastly, citizen groups in cities around the world are advancing new forms of "governance through activism" that represent "an attempt to use collective action to construct alternatives to promote different decisions and management of the environment" (Castán Broto 2019: 165). In the context of growing climate apartheid, such actions are necessary for introducing a more transformative perspective on climate urbanism amid growing climate apartheid.

3.5 Conclusion

In this chapter, we introduce the concept of climate urbanism as one largely governed by a hegemonic, western narrative of development. Through the prioritization and finance of major physical and digital infrastructure projects, securitization, and carbon control, the current policy paradigm of climate urbanism advances a technocratic, neoliberal approach to development. Such a trajectory has significant potential to exacerbate patterns of exclusion and segregation in the urban landscape. Of greater concern is the way this may materialize amid projected climate-induced crises of large-scale displacement, increased xenophobia, climate retreat, and resource competition. Moreover, the philosophy of climate apartheid transcends scale. Because the dominant mentality of climate urbanism is embedded within historical structures of racial capitalism, heteropatriarchy, and coloniality, it holds the potential to reinvigorate a philosophy of apartheid that uses the climate crisis to justify segregation of the climate privileged and the climate precarious at any level. This means that climate apartheid not only reinforces divisions and segregation *within* cities, but in an era of increased protectionism and climate migration control, it creates new tensions and divisions *among* cities. Furthermore, as Paprocki (2020: 2) has recently noted, such responses to climate change introduce new tensions between rural and urban populations as well, "requir[ing] the active destruction of rural futures in order to forge new, resilient and prosperous urban ones."

These seemingly path-dependent outcomes of climate urbanism are not predetermined, however. As some scholars have suggested, engagements with climate urbanism are not homogenous. Instead, as Robin and Castán Broto (forthcoming: 6) have argued, various modalities of climate urbanism are "[directly challenging] the hegemonic discourses that dominate technocratic or finance-led approaches to deliver low carbon, resilient cities." In doing so, these practices embody Brand's (2018: 18) reminder of the importance of recasting urban futures from the situated gaze of local communities:

> Analytically elevating the plurality and complexity of urban space, shaped through processes of attachment, memory, resistance and racialization brings forth richer substantive knowledge from which development paradigms can be challenged.

These visions and strategic interventions are by no means uniformly progressive, and at times they are contradictory. Yet they introduce alternative pathways to climate urbanism that may secure radically different outcomes. In short, while we maintain ongoing critical engagement with the trajectory of hegemonic climate urbanism and its implications for climate apartheid, it is fundamentally necessary to explore, imagine, and advance perspectives on climate urbanism that move us toward a more just and transformative climate-changed future.

References

Ajibade, I. (2017). Can a future city enhance urban resilience and sustainability? A political ecology analysis of Eko Atlantic city, Nigeria. *International Journal of Disaster Risk Reduction, 26*, 85–92.

Bond, P. (2016). Who wins from "Climate Apartheid?": African climate justice narratives about the Paris COP21. *New Politics, 15*(4), 83.

Brand, A. L. (2018). The duality of space: The built world of Du Bois' double-consciousness. *Environment and Planning D: Society and Space, 36*(1), 3–22.

Brisman, A., South, N., & Walters, R. (2018). Southernizing green criminology: Human dislocation, environmental injustice and climate apartheid. *Justice, Power and Resistance, 2*(1), 1–21.

Buckingham, S., & Le Masson, V. (Eds.). (2017). *Understanding climate change through gender relations*. London: Routledge.

Castán Broto, V. (2019). *Urban energy landscapes*. Cambridge: Cambridge University Press.

Chaturvedi, S., & Doyle, T. (2015). *Climate terror: A critical geopolitics of climate change*. Basingstoke: Palgrave Macmillan.

Checker, M. (2011). Wiped out by the "greenwave": Environmental gentrification and the paradoxical politics of urban sustainability. *City & Society, 23*(2), 210–229.

Datta, A., & Shaban, A. (Eds.). (2016). *Mega-Urbanization in the global south: Fast cities and new urban utopias of the postcolonial state*. London: Taylor and Francis.

Davoudi, S. (2014). Climate change, securitisation of nature, and resilient urbanism. *Environment and Planning C: Government and Policy, 32*(2), 360–375.

Dawson, A. (2017). *Extreme cities: The peril and promise of urban life in the age of climate change*. Brooklyn, NY: Verso Books.

Demissie, F. (2007). Imperial legacies and postcolonial predicaments: An introduction. *African Identities, 5*(2), 155–165.

Dobbs, R., Smit, S., Remes, J., Manyika, J., Roxburgh, C., & Restrepo, A. (2011). Urban World: Mapping the economic power of cities. McKinsey Global Institute. Retrieved from https://www.mckinsey.com/featured-insights/urbanization/urban-world-mapping-the-economic-power-of-cities.

Dotson, K. (2018). On the way to decolonization in a settler colony: Re-introducing Black feminist identity politics. *AlterNative: An International Journal of Indigenous Peoples, 14*(3), 190–199.

DuPuis, E. M., & Greenberg, M. (2019). The right to the resilient city: Progressive politics and the green growth machine in New York City. *Journal of Environmental Studies and Sciences, 9*(3), 352–363.

Estrada, F., Botzen, W. J., & Tol, R. S. J. (2017). A global economic assessment of city policies to reduce climate change. *Nature Climate Change, 7*(6), 403–406.

Fields, D., Bissell, D., & Macrorie, R. (2020). Platform methods: studying platform urbanism outside the black box, *Urban Geography, 41*(3), 462–468. https://doi.10.1080/02723638.2020.1730642.

Floater, G., Rode, P., Robert, A., Kennedy, C., Hoornweg, D., Slavcheva, R., & Godfrey, N. (2014). Cities and the New Climate Economy: The transformative role of global urban growth. Retrieved from http://eprints.lse.ac.uk/60775/1/NCE%20Cities%20Paper01.pdf.

Global Commission on Economy and Climate (GCEC). (2016). The sustainable infrastructure imperative: Financing for better growth and development. Retrieved May 24, 2017, from http://www.un.org/pga/71/wp-content/uploads/sites/40/2017/02/New-Climate-Economy-Report-2016-Executive-Summary.pdf.

Fields, D., Bissell, D., & Macrorie, R. (2020). Platform methods: studying platform urbanism outside the black box, Urban Geography, 41(3), 462–468. https://doi.10.1080/02723638.2020.1730642.

Goh, K. (2019). Urban waterscapes: The hydro-politics of flooding in a sinking city. *International Journal of Urban and Regional Research, 43*(2), 250–272.

Hahnhee, H. (2010). Human life in Saemangeum after reclamation. *Journal of the Korean Society for Marine Environment and Energy, 13*(4), 313–326.

Heynen, N. (2016). Urban political ecology II: The abolitionist century. *Progress in Human Geography, 40*(6), 839–845.

Hou, J. (2020). Guerrilla urbanism: Urban design and the practices of resistance. Urban Design International. https://doi.org/10.1057/s41289-020-00118-6

Huq, E., & Mochida, H. (2018). The rise of environmental fascism and the securitization of climate change. Projections. https://doi.org/10.21428/6cb11bd5.

Jacobs, F. (2019). Black feminism and radical planning: New directions for disaster planning research. *Planning Theory, 18*(1), 24–39.

Long, J. (2016). Constructing the narrative of the sustainability fix: Sustainability, social justice and representation in Austin, TX. *Urban Studies, 53*(1), 149–172.

Long, J., & Rice, J. L. (2019). From sustainable urbanism to climate urbanism. *Urban Studies, 56*(5), 992–1008.

Luque-Ayala, A., & Marvin, S. (2015). Developing a critical understanding of smart urbanism? *Urban Studies, 52*(12), 2105–2116.

Moosa, C. S., & Tuana, N. (2014). Mapping a research agenda concerning gender and climate change: A review of the literature. *Hypatia, 29*(3), 677–694.

Norman, E. S. (2017). Standing up for inherent rights: The role of Indigenous-led activism in protecting sacred waters and ways of life. *Society and Natural Resources, 30*(4), 537–553.

Paprocki, K. (2020). The climate change of your desires: Climate migration and imaginaries of urban and rural climate futures. *Environment and Planning D: Society and Space, 38*(2), 248–266.

Paris, C. (2013). *The homes of the super-rich: Multiple residences, hyper-mobility and decoupling of prime residential housing in global cities. In Geographies of the super-rich*. Cheltenham: Edward Elgar Publishing.
Peet, R. (2009). *Unholy trinity: The IMF, World Bank and WTO* (2nd ed.). London: Zed Books Ltd..
Pulido, L., & De Lara, J. (2018). Reimagining 'justice' in environmental justice: Radical ecologies, decolonial thought, and the Black Radical Tradition. *Environment and Planning E: Nature and Space, 1*(1–2), 76–98.
Quastel, N. (2009). Political ecologies of gentrification. *Urban Geography, 30*(7), 694–725.
Ranganathan, M., & Bratman, E. (2019). From urban resilience to abolitionist climate justice in Washington, DC. Antipode, (early view). https://doi.org/10.1111/anti.12555.
Rice, J. L. (2014). Public targets, private choices: Urban climate governance in the Pacific Northwest. *The Professional Geographer, 66*(2), 333–344.
Rice, J. L. (2016). *"The everyday choices we make matter": Urban climate politics and the postpolitics of responsibility and action. Towards a Cultural Politics of Climate Change: Devices, Desires, and Dissent* (pp. 110–126). Cambridge: Cambridge University Press.
Rice, J. L., Cohen, D. A., Long, J., & Jurjevich, J. R. (2020). Contradictions of the climate-friendly city: New perspectives on eco-gentrification and housing justice. *International Journal of Urban and Regional Research, 44*(1), 145–165.
Robin, E., & Castán Broto, V. (forthcoming). Towards a postcolonial perspective on climate urbanism. *International Journal of Urban and Regional Research*.
Robinson, C. J. (2005). *Black Marxism: The making of the black radical tradition*. Chapel Hill: University of North Carolina Press.
Rolnik, R. (2013). Late neoliberalism: The financialization of homeownership and housing rights. *International Journal of Urban and Regional Research, 37*(3), 1058–1066.
Silver, J. (2018). Suffocating cities: Urban political ecology and climate change as social-ecological violence. Urban political ecology in the anthropo-obscene: Interruptions and possibilities, 129–146.
Silver, J. D. (2014). The rise of Afro-Smart cities should be viewed with caution. Africa at LSE. Retrieved from http://eprints.lse.ac.uk/75379/1/Africa%20at%20LSE%20%E2%80%93%20The%20rise%20of%20Afro-Smart%20cities%20should%20be%20viewed%20with%20caution.pdf.

Sultana, F. (2014). Gendering climate change: Geographical insights. *The Professional Geographer, 66*(3), 372–381.

Swyngedouw, E. (2013). The non-political politics of climate change. *ACME: An International E-Journal for Critical Geographies, 12*(1), 1–8.

Theodore, N., Peck, J., & Brenner, N. (2011). Neoliberal urbanism: Cities and the rule of markets. In G. Bridge & S. Watson (Eds.), *The new companion to the city* (pp. 15–25). Oxford: Blackwell-Wiley.

Thomas, K., Hardy, R. D., Lazrus, H., Mendez, M., Orlove, B., Rivera-Collazo, I., et al. (2019). Explaining differential vulnerability to climate change: A social science review. *Wiley Interdisciplinary Reviews: Climate Change, 10*(2), e565.

Tuana, N. (2019). Climate apartheid: The forgetting of race in the anthropocene. *Critical Philosophy of Race, 7*(1), 1–31.

Watts, M. (2017). Cities spearhead climate action. *Nature Climate Change, 7*(8), 537.

Webber, M., Crow-Miller, B., & Rogers, S. (2017). The South–North water transfer project: Remaking the geography of China. *Regional Studies, 51*(3), 370–382.

While, A., Jonas, A. E., & Gibbs, D. (2004). The environment and the entrepreneurial city: Searching for the urban 'sustainability fix' in Manchester and Leeds. *International Journal of Urban and Regional Research, 28*(3), 549–569.

4

The New Climate Urbanism: Old Capitalism with Climate Characteristics

Linda Shi

4.1 Introduction

Everywhere we look, pronouncements from politicians, city networks, and research studies tell us that cities are both the greatest sinners against the climate and seeds of global ecological salvation. From 100 Resilient Cities to 10,000 signatories of the Global Covenant of Mayors for Climate and Energy, cities are increasingly portrayed as the primary agents for advancing societal decarbonization and reducing climate vulnerability. Reflecting this discourse as well as developments on the ground, this edited volume articulates a "New Climate Urbanism"—especially in the introduction by Castán Broto, Robin, and While, and the chapter by Long, Rice, and Levenda. While cities are indeed adopting climate actions that are reshaping urban socio-spatial development, I argue in this chapter that emerging urban responses to climate change are not necessarily new or urban. Rather, competitive dynamics of

L. Shi (✉)
Cornell University, Ithaca, NY, USA
e-mail: lindashi@cornell.edu

V. Castán Broto et al. (eds.), *Climate Urbanism*,
https://doi.org/10.1007/978-3-030-53386-1_4

adaptation within and between metropolitan regions appear to recapitulate capitalist processes reproducing spatial, social, and ecological inequality. Privileging cities as the primary or sole actors involved in urban climate action can overlook historic multi-scalar and multi-level contestations that drive changes at the city scale. Naomi Klein in *This Changes Everything* (2014: 459) writes:

> as the furthest-reaching crisis created by the extractivist worldview, and one that puts humanity on a firm and unyielding deadline, climate change can be the force—the grand push—that will bring together all of these still living [social, labor, and environmental] movements.

However, this chapter shows that from inter-city competition over climate-resilient development in US cities to urban-rural competition over natural resource governance in Asian cities, climate urbanism to date suggests not so much a transformative "grand push" but a Freudian death-drive capitalism with climate characteristics. If this is not what most people think climate urbanism "should be," then it falls on all of us to ask what kinds of urbanism might be more fruitful. The chapter concludes with some reflections and the remainder of the book features contributions that identify seeds of this greater ambition.

4.2 Metropolitan (Urban-Urban) Dynamics of Exclusionary Resilience

Climate change tightens the screws of competitive urbanism. In an age of globalization, fiscal austerity, and decentralization, cities sustain themselves through entrepreneurial competition regionally, nationally, and internationally. Climate change injects an unprecedented shock to this system by squeezing cities of available water resources and habitable land and increasing costs of development, service provision, and disaster recovery. Market responses to climate risks—such as rising insurance premiums, realignment of mortgage markets and property valuation in coastal areas, and falling bond ratings for at-risk municipal or other government bonds—are already or will inject volatility into the bases of municipal fiscal function (Taylor 2020). Evolving research not only identifies the overall potential for financial loss, but the implications for fiscal vulnerability at the city scale (Miao et al. 2018; Shi and Varuzzo 2020; UCS 2018). Cities within and between metropolitan regions compete with one another both to survive the impacts of climate change and to use climate resilience as a narrative and marketing device (see Chaps. 3, 8, and 9).

The experiences of metropolitan efforts to coordinate city-level adaptation to date is indicative of a competitive approach to climate urbanism. As growing numbers of cities began to adapt, many practitioners began to recognize the problems of urban climate adaptation planning (Adams et al., 2016; Shi 2019). For many, barriers to digesting climate projections, undertaking vulnerability assessments, and adopting climate adaptation plans are too onerous (Hamin et al. 2014; Moser and Ekstrom 2012; Shi et al. 2015; see also Chap. 10). Others lament that cities cannot control decisions by utilities, highways, airports, state regulations, nor easily advance adaptation projects given multiple, inconsistent permitting standards (Mukheibir et al. 2013; Vedeld et al. 2016). As some cities began to elevate seawalls or roads, practitioners foresaw a mess of uncoordinated infrastructure systems and the potential for spillover effects. In response, a growing number of metropolitan regions in Europe, Canada, and the United States have launched metropolitan-scale collaborations for climate adaptation (e.g. Adams and French 2019; Bauer and Steurer 2014; Vella et al. 2016). These initiatives aim to catalyse adaptation across metropolitan regions to achieve systems-wide resilience. My research of five such collaboratives in the United States finds that they have helped to build awareness, equalize access to high-quality climate data and training, and lobby for resources (Shi 2019). Recent design competitions—New York's Rebuild by Design (2013); the Changing Course design competition for the Lower Mississippi River Delta (2014); San Francisco Rebuild by Design (2018)—have also enlivened the public's ecological imagination as many nature-based proposals followed ecological rather than administrative boundaries.

Yet, despite efforts to advance collective, equitable, metropolitan-scale adaptation, local implementation remains highly uneven and unequal. Cities that can attract investment have doubled down on waterfront development, often contradicting their own climate vulnerability assessments and regional frameworks (Olazabal et al. 2019; Berke et al. 2015; Shi and Varuzzo 2020; Woodruff and Stults 2016). Wealthier, high-risk municipalities have reconciled this contradiction by pushing for climate-resilient designs particularly in high-end developments, such as elevated buildings, wash-through first floors, wind shear resistance, and elevated HVAC systems, as well as building elevated parks like Hunter's Point and the East River in New York City. The City of Miami Beach and Miami-Dade County are each raising $100 million or more (from water utility charges and bonds respectively) to fund road elevation and drainage upgrades. On the other hand, smaller suburban municipalities and those in "down markets" lack the funding for implementation, easily redevelopable sites from which to squeeze new revenues, and wealthy residents who can afford to pay more for services. An adaptation gap is emerging within and across metropolitan areas as those who can adapt invest in upgrades and armouring infrastructure, and those who cannot afford to adapt fear becoming climate slums (Eubanks 2016).

Competition is also emerging within and between metropolitan regions for status as the preferred "receiving destinations" for people migrating out of flood-prone areas. After Hurricane Katrina, roughly 100,000 people left New Orleans for Baton Rouge, Houston, Dallas, Atlanta, Tampa, and other cities. Studies expect future disaster events to continue to drive internal migration (Hauer 2017). In flood-prone "sending areas," markets are beginning to see depressed property value appreciation or actual value loss (Bernstein et al. 2019; McAlpine and Porter 2018). By contrast, financial, insurance, and development industries are racing to model which areas are the best sites in which to invest in the future (Keenan 2019). Real estate investors are beginning to incorporate climate impacts into their portfolios, although most still rely on insurance to mitigate risks and there is not yet a measurable change to their investment practices (ULI and Heitman 2018). In Southeast Florida, the ground zero of sea level rise in the United States, unincorporated parts of Miami-Dade County on higher land have seen land values grow faster than low-lying places like Miami Beach (Keenan et al. 2018). Low-income communities of colour on higher ground like Little Haiti in the City of Miami have seen rising demand for property in their neighbourhoods. Taking a page from China's Belt and Road Initiative, political scientists Dolšak and Prakash (2019) have proposed a "Green Rustbelt and Road Initiative" to revitalize post-industrial cities of the American Northeast and Midwest with new green industries and climate displaced persons. Whether market shifts and individual choices bring investment to long-neglected geographies with underutilized infrastructure or add housing and transportation pressures to rapidly growing inland metropolises, marginalized communities who have for decades struggled to sustain a right to the city will be among the most impacted.

These emerging dynamics of intra- and inter-metropolitan responses to climate change show that "New Climate Urbanism" can be speculative, competitive, and exclusionary, as described in several contributions across this book. Speculative because individuals and corporations are eyeing future receiving cities, speculatively purchasing land, displacing residents, and anticipating new rounds of a spatial fix (Harvey 2001; Taylor 2020). Competitive because cities ultimately respond to their own residents, budgets, and regulatory requirements, and act in their own self-interest (see Chap. 9). Despite efforts to advance collective action, they necessarily plan and implement adaptation strategies individually and compete for government funding and private development, resulting in short-term decisions that may produce spillover effects to their neighbours. Exclusionary because so much of the emphasis thus far has been on design and infrastructure fixes that are expensive to build and therefore tied to higher-end developments that justify the expense (see Chap. 8). Even where cities espouse nature-based solutions, such efforts can result in gentrification and exclusion (Wachsmuth and Angelo 2018; Anguelovski et al. 2019; Curran and Hamilton 2017; Dooling 2009). Administrative fragmentation, long a factor in racial and class segregation, now produces new dynamics of spatial injustice (Hodson and Marvin 2009).

4.3 Territorial (Urban-Rural) Dynamics of Extractive Resilience

Securitizing cities under climate change also entails landscape-scale efforts to ensure that cities have enough water, electricity, and flood protection. Cities are extractive and dependent on "rural hinterlands" for basic resources—water, energy, food, and natural resources that supply commodities (Cronon 1992; Duarte-Abadía and Boelens 2019; Hidalgo-Bastidas and Boelens 2019; Kaika and Swyngedouw 2011; Silver 2015). The impacts of climate change on both rural and urban areas inject new volatility into what historian William Cronon calls the "elaborate urban-rural hierarchy" (Cronon 1992: 378). This requires climate urbanism research to look at dynamics of appropriation, enclosure, and extraction within and beyond city boundaries. In a recent study, my students and I explore how climate change and efforts to adapt water management in four Asian metropolitan regions affect surrounding rural communities (Shi et al. under review). We developed in-depth case studies of water supply management in the regions around Mumbai (India) and Khulna (Bangladesh), and floodwater management around Bangkok (Thailand) and Metro Manila (the Philippines). Examining these cases together highlights how climate change impacts existing urban-rural dynamics across the spectrum of water management.

Efforts to ensure the resilience of these four metropolitan centres treat rural areas as sacrifice zones. Mumbai and Khulna are conscripting water from rural areas for urban domestic consumption rather than addressing problematic internal water allocations or inefficiencies. Bangkok has built and continues to enlarge rings of levees that protect the capital from floods while flooding rural areas, even as the city ignores its own land use plans and permits development on floodplains reserved for floodwater retention. In Metro Manila, governments relocate urban informal settlers from along drainage ways to rural provinces to improve drainage in the capital, while ignoring the impacts to the floodplain by middle- and upper-class development and historic inadequate housing construction for the urban poor. The outcomes of urban adaptation actions often compound growing climate vulnerability in rural areas, themselves struggling with changing hydrological patterns, among other social and developmental changes. Rural precarity then drives more urban migration. Cities' greater political power and bureaucratic capacity, backed by subnational, national, and international development ambitions, tend to overwhelm rural resistance. In turn, unjust privileging of urban resilience contributes to intensifying urban-rural conflicts. In Bangkok, residents of rural provinces broke 14 water gates following months of inundation, and both water projects in Mumbai and Khulna experienced years of lawsuits. The fragmentation of governance institutions and administrative entities across these large geographic scales challenges efforts to coordinate or rationalize development and water use. These institutional uncertainties further enable dominant political forces to grab resources at the expense of others.

These dynamics suggest, firstly, that cities are not only unintentionally causing spillover effects in pursuit of their own well-being through climate responses, but also privileging their resilience over that of the "other." These actions go beyond the exclusionary to the actively extractive, as cities seek additional water or floodable land from other geographies. More empowered groups mount a resistance such that the most vulnerable groups—informal settlers without land tenure, indigenous tribes, and subsistence fishing and farming communities—that cannot legally or politically resist urban projects are ultimately the ones most impacted. Their resilience and well-being under climate change are never part of assessments, revealing the double standard for whose resilience really counts. Cities' growing prominence as climate actors and their search for investments to build climate-proof enclaves within their boundaries reinforces this incongruity (Chap. 3). Depriving rural areas of fundamental land and water resources denies their ability to survive and sustain the social, governance, and ecological relationships that historically stewarded natural landscapes.

Second, these cases demonstrate that urban responses to climate change implicate geographies far beyond their immediate borders. Cities leading climate action are often represented as dots on a map, as if each dot captures the totality of that geography and all dots are equal and harmonious to others on the map. In reality, of course, these dots lie within metropolitan regions including dozens or hundreds of cities that exist as individual jurisdictions precisely because they value their autonomy. Brenner and Schmid argue the definition of "urban" is a chaotic concept and an artefact of statistics. Instead, they call for "a new conceptualization of urbanization processes both within and beyond those settlement spaces that are demarcated as 'cities'" (2014: 749). Others critique academics for similarly committing "methodological cityism" (Angelo and Wachsmuth 2015) by focusing almost exclusively on the local as the relevant scale of climate adaptation (Nalau et al. 2015; Preston et al. 2013; Chap. 5). While cities do play a lead role in the United States, in the context of centralized Asian countries, these projects are more often the result of decisions at higher levels of government than of local actors.

The scale suggested by "New Climate Urbanism" belies the multi-scalar dimensions of "the urban" (see also Chap. 7). A focus on *cities* creates dichotomous geographies of urban and non-urban, while neglecting relationships that connect these geographies (Roy 2005). A focus on the *urban* can overlook the reality of administrative boundaries that divide a seemingly contiguous urban field into hundreds of fiefdoms with uneven capacity and authority. While the examples given above reference integrated landscapes like metropolitan regions or watersheds, the scale of urban climate securitization reflects the global reach of contemporary capitalism and urbanism (Moser and Hart 2015). Gulf countries, for example, have purchased large tracts of land in East Africa to secure food production given their own lack of water and arable land (Todman 2018). Rising nationalism and efforts to restrict immigration into Europe and the United States in turn reflect national responses to secure nations from climate-displaced migrants. At a minimum, scholars and practitioners must be cognizant of these alternative geographies in assessing urban climate action. A critical task for scholars is to put forth workable alternatives to "cities" that facilitate assessment, planning, governance, and decision-making. It is likely that no such scalar metric exists universally, which itself frustrates efforts to scale up equitable adaptation.

4.4 Conclusion: Imagining Alternative Climate Urbanisms

Cities that have the political and economic clout are enacting a capitalist endgame that rehashes and intensifies existing dynamics of urban developmentalism, social and environmental exploitation, and deepening inequality (Anguelovski et al. 2019; Dooling 2009; Long and Rice 2019). In coastal areas, cities are pragmatically, if cynically, maximizing coastal development to reap whatever revenues and profits they can before the music stops. Other urban centres are wresting resources from rural hinterlands and disposing unwanted urban poor residents to rural peripheries, long-standing practices now buoyed by the added justification of climate change. If this is an era of New Climate Urbanism, these examples suggest that urban responses to climate threats are not necessarily new or even urban. Many of the underlying drivers—fiscalization and monetization of land value; enclosure of the commons and eradication of people from land; boundary-setting to control who can access resources; capital mobility and hyper-competition among municipalities—are depressingly familiar. This is business-as-usual capitalism with climate characteristics. How can scholars, funders, activists, and practitioners challenge these tendencies? Below, I highlight three potential shifts in adaptation practice that can help redirect the growing global industry around adaptation and resilience planning.

Expanding the geographic and temporal scale of resilience planning. There is an urgent need to reframe global resilience discourse to encompass alternative geographic and temporal scales. For one, dominant global networks raise up exemplary cities in ways that elide the complicated territorial infrastructure, ecological, and political systems in which cities sit. From a management perspective, ecosystem scales such as watersheds offer a logical scale at which to manage the hydrological impacts of climate change. The scale of metropolitan (even subnational) regions encompassing urban areas and rural hinterlands also enables more holistic approaches to study the impacts of climate change, the politics of land and resource governance institutions, and identify collective, transformative adaptation strategies. The lack of governance institutions in most countries to manage crises at the regional scale underscores the need to start conversations on territorial governance. The Covid-19 pandemic is reviving interest in relocalization to mitigate vulnerabilities associated with global integration. This potentially awakens opportunities to consider socially and ecologically restorative approaches to regional food, natural resource, and development planning.

For another, the future orientation of climate projections has produced adaptation plans that often forget and erase the historic unsustainable and unjust development processes that have resulted in present-day vulnerabilities. Expanding the historic perspective of resilience initiatives forces adaptation planning to grapple with drivers of urbanization, past processes of resource extraction, pollution, and depletion, and these impacts on different identity groups. These are uncomfortable issues involving conflicts and trade-offs but which are necessary to confront to advance towards transformative change beyond incremental "win-win" solutions.

Looking to the margins rather than the global hubs of urbanization for inspiration. The inspiration for countering dominant, regressive strains of adaptation is unlikely to come from global cities and their governing elites because of their ability to securitize existing centres of wealth and power. As this chapter demonstrates, it is the most politically, economically, and culturally powerful cities that can exclude and extract in seeking resilience. From Jakarta's Great Garuda to Lagos' Eko Atlantis to Rotterdam's Room for the River to New York City's BIG U, these cities have incomparable ability to raise funds, attract investment, and launch large-scale infrastructure projects. In these cities, how are sacrificed and overlooked neighbourhoods organizing their own adaptation? How are secondary, tertiary, suburban, peri-urban, or semi-rural municipalities that lack capacity to plan or invest in infrastructure-based adaption responding? How are indigenous communities whose lifeways entail living with ecological flux dealing with these changes, given their reliance on ecosystems for livelihoods? These "views from off the map" (Robinson 2002) are likely to reveal alternative adaptation strategies requiring collective action and non-infrastructural solutions by dint of necessity or alternative values.

There is an opportunity to uplift alternative perspectives of "resilience" from those who have always been outside of or excluded from the benefits of capitalist urbanization. The planning field increasingly accounts for social vulnerability, equity, and justice in adaptation and resilience. While an improvement from mainstream resilience planning, this typically entails prioritizing vulnerable communities for emergency aid and disaster risk reduction. Instead, social scientists can help synthesize case studies of indigenous knowledge, community-based adaptation, and small-medium city adaptation efforts and identify clear implications for policymakers. The Earth Systems Governance Project has worked to build a global community of scholars over the last decade and advance knowledge synthesis and harvesting (Biermann et al. 2019). This and other efforts must work to ensure this knowledge gains the equivalent salience and recognition as C40 and 100 Resilient Cities.

Connecting and empowering local and transnational social movements. Globally, social movements on the left and right are intensifying, from Extinction Rebellion to Black Lives Matter to the rise of populism and nationalism worldwide. These movements are deeply divided by ideology, identity, and geography, yet share undercurrents of opposition to globalization, income and spatial inequality, and environmental decline. One critical need is for movement alignment that can build towards a bigger "we" to counteract dynamics of competitive and extractive urbanism under resource scarcity. An added challenge is bridging activism with communities of practice proposing and debating specific institutional reforms. Property rights and land tenure regimes, fiscal policy, risk mitigation instruments like insurance, and mortgage-lending policies structure the politics around development decisions. Community activism seeks remedies to immediate threats but rarely offers specific implementable reforms to broader institutions shaping urban development. Instead, highly technical professional experts debate policy changes in these domains, bypassing the altogether separate sphere of activism. There is an opportunity to train both communities of practice to speak one another's language and enable more creative and productive dialogue, learning, and transformation.

Depression, grief, and a sense of helplessness are occupational hazards for natural and social scientists researching climate change. But I reject the notion that critical scholars exist to document the demise of society in an age of climate change. What is the value academia to broader society if not to give hope of better paths forward (see also Chap. 2)? Scholars play a key role in the above proposals. They can support alternative urbanisms by synthesizing knowledge, such as the innumerable case studies we conduct, to show clear alternative pathways. Researchers can become, train, or work with boundary actors to bridge the worlds of activism and institutional governance. As keepers of disciplinary knowledge, we can inform and transform pedagogy and academic or professional institutions that define meaning and norms. Scholars are excellent at identifying and describing the problems with the world; now we must transform our own practice to support a "grand push" towards more liberating climate futures.

References

Adams, S., Crowley, M., Forinash, C., & McKay, H. (2016). *Regional Governance for Climate Action*. Montpelier, VT: Institute for Sustainable Communities.

Adams, S., & French, K. (2019). *Regional Collaboratives for Climate Change – A State of the Art*. Montpelier, VT: Institute for Sustainable Communities.

Angelo, H., & Wachsmuth, D. (2015). Urbanizing urban political ecology: A critique of methodological cityism: Urbanizing urban political ecology. *International Journal of Urban and Regional Research, 39*(1), 16–27. https://doi.org/10.1111/1468-2427.12105.

Anguelovski, I., Connolly, J. J. T., Pearsall, H., Shokry, G., Checker, M., Maantay, J., et al. (2019). Opinion: Why green "climate gentrification" threatens poor and vulnerable populations. *Proceedings of the National Academy of Sciences, 116*(52), 26139–26143. https://doi.org/10.1073/pnas.1920490117.

Bauer, A., & Steurer, R. (2014). Multi-level governance of climate change adaptation through regional partnerships in Canada and England. *Geoforum, 51*, 121–129.

Bernstein, A., Gustafson, M., & Lewis, R. (2019). Disaster on the Horizon: The price effect of sea level rise. *Journal of Financial Economics, 134*(2), 253–272.

Berke, P., Newman, G., Lee, J., Combs, T., Kolosna, C., & Salvesen, D. (2015). Evaluation of networks of plans and vulnerability to hazards and climate change: A resilience scorecard. *Journal of the American Planning Association, 81*(4), 287–302. https://doi.org/10.1080/01944363.2015.1093954.

Biermann, F., Betsill, M. M., Burch, S., Dryzek, J., Gordon, C., Gupta, A., Gupta, J., Inoue, C., Kalfagianni, A., Kanie, N., Olsson, L., Persson, Å., Schroeder, H., & Scobie, M. (2019). The earth system governance project as a network organization: A critical assessment after ten years. *Current Opinion in Environmental Sustainability, 39*, 17–23. https://doi.org/10.1016/j.cosust.2019.04.004.

Brenner, N., & Schmid, C. (2014). The 'Urban Age' in Question. *International Journal of Urban and Regional Research, 38*(3), 731–755.

Cronon, W. (1992). *Nature's Metropolis: Chicago and the Great West*. New York: W. W. Norton and Co..

Curran, W., & Hamilton, T. (Eds.). (2017). *Just green enough: Urban development and environmental gentrification*. London: Routledge.

Dolšak, N., & Prakash, A. (2019). Jobs And Climate Change: America's (Rust) Belt And Road Initiative. Forbes. Retrieved from https://www.forbes.com/

sites/prakashdolsak/2019/07/14/jobs-and-climate-change-americas-rust-belt-and-road-initiative.
Dooling, S. (2009). Ecological gentrification: A research agenda exploring justice in the city. *International Journal of Urban and Regional Research, 33*(3), 621–639.
Duarte-Abadía, B., & Boelens, R. (2019). Colonizing rural waters: The politics of hydro-territorial transformation in the Guadalhorce Valley, Málaga, Spain. *Water International, 44*(2), 148–168.
Eubanks, V. (2016). My drowning city is a Harbinger of climate slums to come. The Nation. https://www.thenation.com/article/low-water-mark/.
Hamin, E. M., Gurran, N., & Emlinger, A. M. (2014). Barriers to municipal climate adaptation: Examples from coastal Massachusetts' smaller cities and towns. *Journal of the American Planning Association, 80*(2), 110–122.
Harvey, D. (2001). Globalization and the spatial fix. *Geographische Revue, 2*, 23–30.
Hauer, M. E. (2017). Migration induced by sea-level rise could reshape the US population landscape. *Nature Climate Change, 7*(5), 321–325.
Hidalgo-Bastidas, J. P., & Boelens, R. (2019). The political construction and fixing of water overabundance: Rural–urban flood-risk politics in coastal Ecuador. *Water International, 44*(2), 169–187.
Hodson, M., & Marvin, S. (2009). "Urban ecological security": A new urban paradigm? *International Journal of Urban and Regional Research, 33*(1), 193–215.
Kaika, M., & Swyngedouw, E. (2011). The urbanization of nature: Great promises, impasse, and new beginnings. In G. Bridge & S. Watson (Eds.), *The new blackwell companion to the city*. Malden, MA: Wiley-Blackwell.
Keenan, J. M. (2019). A climate intelligence arms race in financial markets. *Science, 365*(6459), 1240–1243.
Keenan, J. M., Hill, T., & Gumber, A. (2018). Climate gentrification: From theory to empiricism in Miami-Dade County, Florida. *Environmental Research Letters, 13*(5).
Klein, N. (2014). This Changes Everything: Capitalism vs. The Climate. Simon and Schuster.
Long, J., & Rice, J. L. (2019). From sustainable urbanism to climate urbanism. *Urban Studies, 56*(5), 992–1008. https://doi.org/10.1177/0042098018770846.

McAlpine, S. A., & Porter, J. R. (2018). Estimating recent local impacts of sea-level rise on current real-estate losses: A housing market case study in Miami-Dade, Florida. *Population Research and Policy Review, 37*(6), 871–895.
Miao, Q., Hou, Y., & Abrigo, M. (2018). Measuring the financial shocks of natural disasters: A panel study of U.S. States. *National Tax Journal, 71*(1), 11–44.
Moser, S. C., & Ekstrom, J. A. (2012). Identifying and overcoming barriers to climate change adaptation in San Francisco bay: Results from case studies [White Paper]. California Energy Commission.
Moser, S. C., & Hart, J. A. F. (2015). The long arm of climate change: Societal teleconnections and the future of climate change impacts studies. *Climatic Change, 129*(1–2), 13–26.
Mukheibir, P., Kuruppu, N., Gero, A., & Herriman, J. (2013). Overcoming cross-scale challenges to climate change adaptation for local government: A focus on Australia. *Climatic Change, 121*(2), 271–283.
Nalau, J., Preston, B. L., & Malo, M. C. (2015). Is adaptation a local responsibility. *Environmental Science and Policy, 48*, 89–98.
Olazabal, M., Gopegui, M. R., de, Tompkins, E. L., Venner, K., & Smith, R. (2019). A cross-scale worldwide analysis of coastal adaptation planning. *Environmental Research Letters, 14*(12), 124056. https://doi.org/10.1088/1748-9326/ab5532.
Preston, B. L., Mustelin, J., & Maloney, M. C. (2013). Climate adaptation Heuristics and the science/policy divide. *Mitigation and Adaptation Strategies for Global Change, 20*(3), 467–497.
Robinson, J. (2002). Global and world cities: A view from off the map. *International Journal of Urban and Regional Research, 26*(3), 531–554.
Roy, A. (2005). Urban informality: Toward an epistemology of planning. *Journal of the American Planning Association, 71*(2), 147–158.
Silver, J. (2015). Disrupted infrastructures: An urban political ecology of interrupted electricity in accra. *International Journal Urban Regional, 39*, 984–1003. https://doi.org/10.1111/1468-2427.12317.
Shi, L. (2019). Promise and paradox of metropolitan regional climate adaptation. *Environmental Science and Policy, 92*, 262–274.
Shi, L., Chu, E., & Debats, J. (2015). Explaining progress in climate adaptation planning across 156 U.S. municipalities. *Journal of the American Planning Association, 81*(3), 191–202.
Shi, L., & Varuzzo, A. M. (2020). Rising seas, surging fiscal stress: Patterns and implications of municipal fiscal vulnerability to climate change in Massachusetts. *Cities, 100*, 102658.

Shi, L., Ahmad, F., Shukla, P., & Yupho, S. (under review). Urban security, rural sacrifice: Governing water in an age of climate change. Global Environmental Change.

Taylor, Z. J. (2020). The real estate risk fix: Residential insurance-linked securitization in the Florida metropolis: Environment and Planning A: Economy and Space. Early view: https://doi.org/10.1177/0308518X19896579.

Todman, W. (2018). The gulf scramble for Africa: GCC states' foreign policy laboratory [CSIS Brief]. Center for Strategic and International Studies. https://www.csis.org/analysis/gulf-scramble-africa-gcc-states-foreign-policy-laboratory.

UCS. (2018). Underwater: Rising seas, chronic floods, and the implications for US coastal real estate (p. 28). Union of Concerned Scientists.

ULI and Heitman. (2018). Future proofing real estate from climate risks. Urban Land Institute and Heitman Real Estate Management Firm. http://www.heitman.com/wp-content/uploads/2018/10/Future-Proofing-Real-Estate.pdf.

Vedeld, T., Coly, A., Ndour, N. M., & Hellevik, S. (2016). Climate adaptation at what scale? Multi-level governance, resilience, and coproduction in Saint Louis, Senegal. *Natural Hazards, 82*(2), 173–199. https://doi.org/10.1007/s11069-015-1875-7.

Vella, K., Butler, W. H., Sipe, N., Chapin, T., & Murley, J. (2016). Voluntary collaboration for adaptive governance: The Southeast Florida regional climate change compact. *Journal of Planning Education and Research, 36*(3), 363–376.

Wachsmuth, D., & Angelo, H. (2018). Green and gray: New ideologies of nature in urban sustainability policy. *Annals of the American Association of Geographers, 108*(4), 1038–1056. https://doi.org/10.1080/24694452.2017.1417819.

Woodruff, S. C., & Stults, M. (2016). Numerous strategies but limited implementation guidance in US local adaptation plans. *Nature Climate Change, 6*, 796–802.

5

Understanding the Governance of a New Climate Urbanism

Sirkku Juhola

5.1 Introduction

The emergence of climate change as a result of urbanization and possible threat to urban areas has prompted a significant effort both in terms of scholarship and practical action in cities. With more than half of the world population now living in cities, we need to understand how cities are governed. In relation to climate change, this means understanding whether and how mitigation and adaptation strategies shape the way cities develop, how urban governance is changing as a result, and how these changes manifest in urban decision-making. When discussing the "climate" in climate urbanism, it is useful to keep in mind that climate policy consists of both mitigation and adaptation strategies. Mitigation relates to the reduction of greenhouse gas emissions, while adaptation relates to actions seeking to reduce climate risks or take advantage of climate

S. Juhola (✉)
Ecosystems and Environment Research Programme, University of Helsinki and Helsinki Institute of Sustainability Science (HELSUS), Helsinki, Finland
e-mail: sirkku.juhola@helsinki.fi

V. Castán Broto et al. (eds.), *Climate Urbanism*,
https://doi.org/10.1007/978-3-030-53386-1_5

impacts. Historically, adaptation has been overlooked and some have argued that a focus on adaptation would reduce interests to mitigate emissions (Pielke et al. 2007). Furthermore, there have been long-standing debates on the differences between adaptation and mitigation as part of integrated climate policy (Tol 2005). According to many—including the IPCC—mitigation efforts benefit the common good as the emissions reductions benefit everyone, irrespective of who acts. Conversely, adaptation is mainly designed to benefit those that implement measures locally, particularly if the action is taken as a response to particular climate risks or impacts that affect a particular place. Thus, trade-offs—but also synergies—may appear between mitigation and adaptation goals as they are most often discussed together with regards to urban governance (Landauer et al. 2015). For instance, trade-offs can play out in terms of physical planning, that is, densification schemes to mitigate emissions from transport may have negative effects on the ability to use nature-based solutions to reduce urban heat island effects or managing surface runoff water (Martens et al. 2009). There will also be inevitable trade-offs in terms of resource allocation and goal-setting for adaptation and mitigation policy, as both compete in the same "environmental" decision-making arena.

Despite these challenges, both adaptation and mitigation strategies will deeply impact the emergence of new forms of climate urbanism by reorienting urban governance and by influencing how the urban form evolves as a result of and in response to climate change. Mitigation efforts for instance reshape how cities consume energy, organize their transport, and provide food. Through adaptation, cities are reconfigured to create safe living environments by reducing risks from extreme weather events and long-term environmental changes. In this chapter, I review existing literature on the governance of adaptation and mitigation in cities to formulate three propositions that can be read as preconditions to understand the emergence of a climate urbanism and its governance. Specifically, I explore how climate issues have been taken up in urban decision-making, how effective these actions are and whether climate actions are a driving force in urban development more broadly. Thus, my first proposition is that to claim that a distinctive climate urbanism is emerging, we need empirical evidence of the institutionalization of climate policy within urban governments, as well as a proliferation of governance arrangements and instruments to engage other actors in urban climate policy. My second proposition is that while there are plenty of climate-related strategies undertaken in cities, often through networks, their actual impact—manifested in changes in the urban form or social life—remains limited. My third proposition explores whether climate policy has become a driving force in urban development, questioning its influence in the broader field of urban policy. In formulating these three propositions, my aim here is to highlight new research directions, and reflect upon the methodological challenges they pose for an emerging research agenda on climate urbanism.

5.2 The Institutionalization of Climate Policy

The first proposition is that to claim that a distinctive climate urbanism is emerging, we need to observe a shift in urban governance as a result of climate change. Indeed, the emergence of urban climate governance has been well-documented, showing that cities have become key actors in addressing climate change and are now using a number of different governance arrangements to accomplish their goals (Betsill and Bulkeley 2006; Bulkeley 2013). However, questions remain on the permanence of these governance arrangements, and on the extent to which these developments permeate across different spheres of governance. Institutionalization here is key. I define institutionalization as the embedding of climate policy objectives into urban governance structure and frameworks with the aim of both mitigating and adapting to climate change across all policy areas. Furthermore, I take inspiration from Aylett (2015) who defines governance as a process of governing within government and beyond government. This implies looking at the multitude of actors engaging in climate policy and recognizing the wide variety of instruments and mechanisms used to adapt to and mitigate against climate change in cities (Anguelovski and Carmin 2011). City governments engage with a variety of other actors across multiple scales to implement climate action, using a multitude of policy instruments.

Two specific issues require attention to trace the emergence of a distinct way of governing climate urbanism. First, it is important to continue to accumulate empirical observations on policy implementation, as also noted by van der Heijden (2019). There is a gap in existing scholarship when it comes to understanding to what extent both mitigation and adaptation policy goals have been adopted as binding urban policy targets. There is an increasing number of global surveys, which have been instrumental in increasing empirical knowledge of local-level implementation (Reckien et al. 2015, 2018; Heidrich et al. 2013; Bassett and Shandas 2010; Guyadeen 2018) of both adaptation (Baker et al. 2012; Araos et al. 2016; Aguiar et al. 2018, Chap. 10) and mitigation strategies (Croci et al. 2017). Many studies are based on policy document analysis, undertaken by an online collection, followed by a translation (if necessary) and a coding of the documents to assess the extent of policy adoption. In some cases, interviews or survey questionnaires are used to supplement this data and confirm the findings. To broadly summarize the above-mentioned reviews, it appears that mitigation is more widely adopted as a policy than adaptation by city governments. For example, about 66% of cities in the European Union have a mitigation plan, but only 26% have an adaptation plan (Reckien et al. 2018). Local authorities that address both policy goals are fewer (Aylett 2015; Reckien et al. 2018). Many of these reviews have been conducted in the global North, where action on climate change overall has been shown to be more prevalent than in the global South (van der Heijden 2019; Chap. 10). In relation to the emergence of climate urbanism, these findings raise questions as to whether climate policy has become institutionalized to the extent that we within the field propose it has.

There are a number of factors that challenge how we research the institutionalization of climate policy in cities worldwide, and how we understand this in relation to climate urbanism. To begin with, obtaining reliable data from cities globally is a challenge and there will always be some form of bias in the framing and methods chosen for any study. There is a selection and confirmation bias that lead us to gravitate towards "good" cases or illustrative cases (van der Heijden 2019; Chap. 9). Thus, while climate policy seems to be institutionalized at the city-level in some parts of the world, there remain immense knowledge gaps with regards to the large number of small- and medium-sized Southern cities that make up this planet and are understudied. There are well-studied front-runner cities (Wolfram et al. 2019), but lack of action and action failures are often not documented. There are also other reasons why methods such as surveys or case studies may not capture all climate-related activities carried out in cities. Policy documents may be vague, unavailable or inaccurate about the actions taken. It is also possible that many city administrations, particularly the front-runner ones, are seeking to exploit multiple policy goals (win-win), and thus many of actions may not be labelled as climate-specific but designated as part of a broader category of sustainability measures. Thus, when coding of policy documents takes place, a number of measures may be left out of the analysis. There are no empirical analyses of this, so it is hard to estimate how much this affects empirical findings.

Another issue relates to the question of actors and instruments in the institutionalization of climate policy. The assumption here is that more actors become engaged in developing and implementing climate policy as it becomes institutionalized into urban governance structures, that is, governance within city governments and beyond city government structures. There are only preliminary global studies on the actors involved, beyond the state, and the modalities of their involvement. For example, an assessment of adaptation efforts for 402 cities globally found that city governments continue to predominantly steer themselves (Klein et al. 2018). The same study also found that the private sector is involved more often than the citizens. The private sector is involved in adaptation efforts mostly through partnerships, while citizens are involved through the provision of information by local governments. As these developments continue, important questions are already being asked about participation, legitimacy, and justice (Bulkeley et al. 2014) and about the politics of urban climate governance more broadly (Hughes 2017). For instance, research has found that questions of who will be adversely affected by adaptation measures are rarely discussed in public forums (Bulkeley et al. 2014). Given the diffused ways in which policy agendas form and are implemented, legitimacy becomes an issue. It is not always clear who has the power to make decisions and who is involved in decision-making processes. There are differences here between mitigation and adaptation in terms of what an equitable outcome might look like. Impacts of mitigation policy and climate impacts both will be experienced unevenly within a city (Rice et al. 2020). These issues have so far been discussed from a global perspective but less so at a local scale (De Cian et al. 2016). This has implications for how different governance arrangements supporting the emergence of climate urbanism will help integrate social and environmental justice concerns in future climate action.

5.3 Implementing Urban Climate Policy

The second proposition states that official urban climate strategies do not necessarily lead to substantive physical and social transformations in cities. There are two main issues with regards to how we understand the concrete implementation of climate policy in cities. The first one is an empirical question of how we can study implementation, and the second one has to do with future research on the topic. On the first point, while the studies discussed in the previous section offer evidence of strategic policy instruments, it is harder to obtain a global picture of the degree of implementation of climate policy on the ground. Rare studies have, for instance, surveyed the implementation of climate experiments in 100 cities (Castán Broto and Bulkeley 2013). Similarly, a review of 124 cities' mitigation plans reveals over 5500 individual mitigation actions (Croci et al. 2017). Policy documents have been used to assess the levels of ambition with regards to implementation: in this study of C40 cities, mitigation and adaptation strategies do not appear to be particularly ambitious, with most of the measures classified as incremental rather than transformational (Heikkinen et al. 2019). Naturally, there are a number of conceptual and methodological challenges in obtaining robust empirical data on the implementation of climate action on a large scale. These challenges are even greater than in assessing institutionalization because implementation can take place over different time scales, ranging from fast-paced experimentation to decades-long transitions in infrastructure, for example. In addition, spatial scales can vary in terms of where the policy is implemented, ranging from building to block and neighbourhood to city-wide level. So far, it appears that there is a tendency to use policy documents as data representing implementation but there is a need to explicitly distinguish between strategy-level policy documents and the actual implementation of measures (Heikkinen et al. 2019; Chap. 10).

There is a fundamental gap in knowledge on the implementation of climate policy over time. There is some evidence that much of climate-related actions, particularly in their initial stages, are driven by externally funded projects for a limited period (Juhola et al. 2012; Chap. 8) with varying degrees of success (Murtinho et al. 2013). Therefore, it is not possible to say how effective many actions have been, to what extent they have fully been implemented and continue to be, and what their implications for urban developments are going forward. This makes it difficult to understand how climate action reshapes urban governance and cities more broadly. On the one hand, the fast pace of policy-making and rapid project lifecycles of new initiatives drive decisions. On the other, there is the slower process of institutionalization and path dependency of policy that works in parallel, also influencing governance structures and the impact of different climate actions (Chap. 7). In some cases, the slowness of implementation may be explained by what van der Heijden (2018) calls the "front runner paradox". According to him, the focus on front-runner cities that innovate and experiment may lead to a situation where the lessons drawn from their experiences are not relevant or do not resonate with the majority of cities. Much of the empirical experiences are collected from front-runner cities where action takes place but their uniqueness may affect the scalability of their initiative. Similarly, the idea of best practices needs further scrutiny. Drawing on the experiences of German cities, Nagorny-Koring (2019) showed that best practices, given their inevitable context dependency, are not always replicable and do not automatically lead to policy change, despite their successful implementation in one specific place.

More broadly, difficulties in tracing implementation and impacts resonate with wider methodological difficulties to assess policy developments comprehensively. Over what time period does one need to observe climate policy developments to understand implementation? To address this "implementation gap" in future research, different research institutes should collaborate to constantly monitor and evaluate policy implementation across cities in different continents over a longer time period. This could involve developing a pilot in a group of cities, which could be expanded through the involvement of climate-oriented city networks, for example. Given the recent nature of both adaptation and mitigation policy fields, there is relatively little follow up or time series-oriented research, assessing or documenting the impact of policies (Fawcett et al. 2017). Naturally, mitigation policies and their effectiveness are measured through reduction in emissions, though not often in real life but with estimates (Dahal and Niemelä 2017; Dahal et al. 2018). There is no such metric available for adaptation (Jurgilevich et al. 2019), where the policy problem itself is too complex and multifaceted to be addressed by a single indicator of success. A broader research question relates to the social impacts of mitigation and adaptation more broadly. As pointed out by Shi et al. (2016), there are very few studies that assess the social implications of successful or unsuccessful implementation of mitigation or adaptation policies over longer periods of time, particularly in relation to equity and fairness. Interesting insights from case studies examining the adaptation planning phases reveal that more inclusive planning processes correspond to higher climate equity and justice outcomes in the short term (Chu et al. 2016). Similarly, planning processes can both protect privileged communities and further disadvantage those already at risk (Anguelovski et al. 2016). This means that some adaptation measures will involve maladaptive outcomes for other actors (Juhola et al. 2016; Chap. 3). Maladaptation has not comprehensively been studied in cities as of yet, but early studies within agriculture, for example, show that most farmers chose measures that will negatively affect them in the future, instead of those that shift vulnerability to other farmers (Neset et al. 2019). These findings mean that we need more research on the actual impact of climate policies in cities to comprehensively assess the transformative and progressive potential of climate urbanism, particularly in relation to social justice.

5.4 Policy Coherence and Competition

The third proposition posits that the emergence of a distinctive climate urbanism would mean that climate policy has become a—if not the—driving force in urban development. This proposition relates to the effectiveness of urban climate policy in achieving its targets within a competitive urban policy environment. After all, it is worthwhile to keep in mind that cities are not built to mitigate and adapt to climate change but to function as liveable cities, and that they are made up of many diverging interests (Chaps. 8 and 9). Thus, when looking at the reconfiguration of urban governance for climate urbanism, one should focus on how climate policy is embedded in urban governance structures (if at all) and how it interacts with other policy domains.

So far, much of the scholarship within urban climate studies focused solely on climate policy, often talking of mainstreaming, policy coherence or comprehensiveness. For example, a recent study examined climate policy integration, that is integration between adaptation and mitigation, in 147 cities, and found that the level of integration was moderate (Grafakos et al. 2020). In their analysis, the authors define integration as a plan that identifies both sources of vulnerability and emissions and systematically considers integration possibilities. The low level of integration, according to the authors, is due to insufficient quantitative evaluation of costs and funding schemes for the comprehensive implementation of adaptation and mitigation measures (Grafakos et al. 2020). Another way of examining the interplay between mitigation and adaptation is to analyse climate policy comprehensiveness. Comprehensiveness, according to Lee and Painter (2015), is manifested in policies that adopt an integrated approach in planning and implementing climate change strategies and mechanisms. Here, the authors construct a series of steps that consist of procedural mitigation steps, sectoral comprehensiveness in mitigation and preparedness for adaptation. Based on a case study of four cities, the authors conclude that those cities that have well-developed governance arrangements are likely to implement a comprehensive climate change policy (Lee and Painter 2015). While these studies highlight the importance of synergies between mitigation and adaptation, neither of them explicitly addresses how climate objectives compete against other urban goals when it comes to urban governance more broadly. Given that mitigation policies target energy use, there is a pressing need to analyse how climate policies compete with other sectoral interests, such as transportation, energy supply and demand, public health, building, waste and water. Similarly, adaptation needs can run counter to the ambitions of land use planning, that is, housing built in flood risk zones.

Thus, while examining the internal coherence of climate policy itself is a worthwhile exercise, I would argue that more attention needs to be paid to sectoral policy coherence within city governance structures and to analyse competing policy agendas more broadly. A study that assesses strategic adaptation actions in three cities pinpoints sources of planning tensions that influence the type adaptation adopted (Chu et al. 2017). Decisions may well be driven by completely different kinds of interests and politics. Thus, while understanding the inner workings and processes of climate policy itself is interesting, it is worth acknowledging that maybe other policy fields are sometimes more important in the context of the city in the first place, such as health care, education, and economic policy (Chap. 9). To quote Lewis Mumford, the city is a "theatre of social action" and a place for "significant collective drama" (Mumford 1961), meaning that climate policy is only a facet of policymaking in any city, and perhaps not even the most significant for emissions and the emergence of vulnerability (e.g. transport and housing strategies might have more effects on resource consumption, GHG emissions and the production of spatially differentiated vulnerabilities). So perhaps climate-oriented researchers, particularly those interested in climate urbanism, should focus not just on mitigation and adaptation initiatives but on other policy areas, particularly those where emissions are generated and climate vulnerabilities (re)produced, in order to understand why and how this happens in the first place. Similarly, adaptation researchers ought to focus more on the processes that create vulnerability and enhance exposure, rather than on how to undertake protective measures (Adelekan 2010; Hardoy and Pandiella 2009). These issues present theoretical and methodological challenges that would need to address how we assess the influence of different policy fields in terms of overall decision-making in cities, and certainly supports the need for more detailed, critical case analyses of the politics of an emerging climate urbanism.

5.5 Conclusion

This chapter discussed the role of governance and policy in supporting the emergence of a distinctive mode of climate urbanism. It highlighted three issue, each of which are linked to how climate policy transforms urban governance. Underlying these three issues I highlighted challenges related to empirical research, despite the significant strides made in the recent past that allow for an examination of the emergence of climate urbanism. In the future, more critical and theoretically diverse research on detailed scales over time is needed, as well as basic empirical research to understand how climate policy transforms how cities are planned, built, and experienced globally.

References

Adelekan, I. O. (2010). Vulnerability of poor urban coastal communities to flooding in Lagos, Nigeria. *Environment and Urbanization, 22*(2), 433–450.

Aguiar, F. C., Bentz, J., Silva, J. M. N., Fonseca, A., Swart, R., Santos, F., & Penha-Lopes, G. (2018). Adaptation to climate change at local level in Europe: An overview. *Environmental Science Policy, 86*, 38–63.

Anguelovski, I., & Carmin, J. (2011). Something borrowed, everything new: Innovation and institutionalization in urban climate governance. *Current Opinion in Environmental Sustainability, 3*(3), 169–175.

Anguelovski, I., Shi, L., Chu, E., Gallagher, D., Goh, K., Lamb, Z., et al. (2016). Equity impacts of urban land use planning for climate adaptation: Critical perspectives from the global north and south. *Journal of Planning Education and Research, 36*(3), 333–348.

Araos, M., Berrang-Ford, L., Ford, J. D., Austin, S. E., Biesbroek, R., & Lesnikowski, A. (2016). Climate change adaptation planning in large cities: A systematic global assessment. *Environmentl Science Policy, 66*, 375–382.

Aylett, A. (2015). Institutionalizing the urban governance of climate change adaptation: Results of an international survey. *Urban Climate, 14*, 4–16.

Baker, I., Peterson, A., Brown, G., & McAlpine, C. (2012). Local government response to the impacts of climate change: An evaluation of local climate adaptation plans. *Landsc ape Urban Plan, 107*, 127–136.

Bassett, E., & Shandas, V. (2010). Innovation and climate action planning: Perspectives from municipal plans. *Journal of the American Planning Association, 76*, 435–450.
Betsill, M. M., & Bulkeley, H. (2006). Cities and the multilevel governance of global climate change. *Global Governance, 12*, 141.
Bulkeley, H. (2013). *Cities and climate change*. Abingdon: Routledge.
Bulkeley, H., Edwards, G. A., & Fuller, S. (2014). Contesting climate justice in the city: Examining politics and practice in urban climate change experiments. *Global Environmental Change, 25*, 31–40. https://doi.org/10.1016/j.gloenvcha.2014.01.009.
Chu, E., Anguelovski, I., & Carmin, J. (2016). Inclusive approaches to urban climate adaptation planning and implementation in the Global South. *Climate Policy, 16*(3), 372–392.
Chu, E., Anguelovski, I., & Roberts, D. (2017). Climate adaptation as strategic urbanism: Assessing opportunities and uncertainties for equity and inclusive development in cities. *Cities, 60*, 378–387.
Croci, E., Lucchitta, B., Janssens-Maenhout, G., Martelli, S., & Molteni, T. (2017). Urban CO2 mitigation strategies under the Covenant of Mayors: An assessment of 124 European cities. *Journal of Cleaner Production, 169*, 161–177.
Dahal, K., Juhola, S., & Niemelä, J. (2018). The role of renewable energy policies for carbon neutrality in Helsinki Metropolitan area. *Sustainable Cities and Society, 40*, 222–232.
Dahal, K., & Niemelä, J. (2017). Cities' greenhouse gas accounting methods: A study of Helsinki, Stockholm, and Copenhagen. *Climate, 5*(2), 31.
De Cian, E., Hof, A. F., Marangoni, G., Tavoni, M., & Van Vuuren, D. P. (2016). Alleviating inequality in climate policy costs: An integrated perspective on mitigation, damage and adaptation. *Environmental Research Letters, 11*(7), 074015.
Fawcett, D., Pearce, T., Ford, J. D., & Archer, L. (2017). Operationalizing longitudinal approaches to climate change vulnerability assessment. *Global Environmental Change, 45*, 79–88.
Grafakos, S., Viero, G., Reckien, D., Trigg, K., Viguie, V., Sudmant, A., et al. (2020). Integration of mitigation and adaptation in urban climate change action plans in Europe: A systematic assessment. *Renewable and Sustainable Energy Reviews, 121*, 109623.

Guyadeen, D. (2018). Do practicing planners value plan quality? Insights from a survey of planning professionals in Ontario, Canada. *Journal of the American Planning Association, 84*, 21–32.

Hardoy, J., & Pandiella, G. (2009). Urban poverty and vulnerability to climate change in Latin America. *Environment and Urbanization, 21*(1), 203–224.

Heidrich, O., Dawson, R., Reckien, D., & Walsh, C. (2013). Assessment of the climate preparedness of 30 urban areas in the UK. *Climatic Change, 120*, 771–784.

Heikkinen, M., Ylä-Anttila, T., & Juhola, S. (2019). Incremental, reformistic or transformational: What kind of change do C40 cities advocate to deal with climate change? *Journal of Environmental Policy and Planning, 21*(1), 90–103.

Hughes, S. (2017). The politics of urban climate change policy: Toward a research agenda. *Urban Affairs Review, 53*(2), 362–380.

Juhola, S., Glaas, E., Linnér, B. O., & Neset, T. S. (2016). Redefining maladaptation. *Environmental Science and Policy, 55*, 135–140.

Juhola, S., Peltonen, L., & Niemi, P. (2012). The ability of Nordic countries to adapt to climate change: Assessing adaptive capacity at the regional level. *Local Environment, 17*(6-7), 717–734.

Jurgilevich, A., Groundstroem, F., Klein, J., Räsänen, A., & Juhola, S. (2019). The emergence and institutionalization of nation adaptation strategies. In E. C. H. Keskitalo & B. L. Preston (Eds.), *Research handbook on climate change adaptation policy*. London: Edward Elgar Publishing.

Klein, J., Araos, M., Karimo, A., Heikkinen, M., Ylä-Anttila, T., & Juhola, S. (2018). The role of the private sector and citizens in urban climate change adaptation: Evidence from a global assessment of large cities. *Global Environmental Change, 53*, 127–136.

Landauer, M., Juhola, S., & Söderholm, M. (2015). Inter-relationships between adaptation and mitigation: a systematic literature review. *Climatic Change, 131*(4), 505–517. https://doi.org/10.1007/s10584-015-1395-1.

Lee, T., & Painter, M. (2015). Comprehensive local climate policy: The role of urban governance. *Urban Climate, 14*, 566–577.

Martens, P., McEvoy, D., & Chang, C. (2009). The climate change challenge: Linking vulnerability, adaptation, and mitigation. *Current Opinion in Environmental Sustainability, 1*(1), 14–18.

Mumford, L. (1961). The city in history: Its origins, its transformations, and its prospects (Vol. 67). Houghton Mifflin Harcourt.

Murtinho, F., Eakin, H., López-Carr, D., & Hayes, T. M. (2013). Does external funding help adaptation? Evidence from community-based water management in the Colombian Andes. *Environmental Management, 52*(5), 1103–1114.

Nagorny-Koring, N. C. (2019). Leading the way with examples and ideas? Governing climate change in German municipalities through best practices. *Journal of Environmental Policy and Planning, 21*(1), 46–60.

Neset, T. S., Wiréhn, L., Klein, N., Käyhkö, J., & Juhola, S. (2019). Maladaptation in Nordic agriculture. *Climate Risk Management, 23*, 78–87.

Pielke, R., Jr., Prins, G., Rayner, S., & Sarewitz, D. (2007). Climate change 2007: Lifting the taboo on adaptation. *Nature, 445*(7128), 597.

Reckien, D., Flacke, J., Olazabal, M., & Heidrich, O. (2015). The influence of drivers and barriers on urban adaptation and mitigation plans—an empirical analysis of European cities. *PLoS One, 10*(8), e0135597.

Reckien, D., Salvia, M., Heidrich, O., Church, J. M., Pietrapertosa, F., De Gregorio-Hurtado, S., D'Alonzo, V., Foley, A., Simoes, S. G., Krkoška Lorencová, E., Orru, H., Orru, K., Wejs, A., Flacke, J., Olazabal, M., Geneletti, D., Feliu, E., Vasilie, S., Nador, C., Krook-Riekkola, A., Matosović, M., Fokaides, P. A., Ioannou, B. I., Flamos, A., Spyridaki, N.-A., Balzan, M. V., Fülöp, O., Paspaldzhiev, I., Grafakos, S., & Dawson, R. (2018). How are cities planning to respond to climate change? Assessment of local climate plans from 885 cities in the EU-28. *Journal of Cleaner Production, 191*, 207–219.

Rice, J. L., Cohen, D. A., Long, J., & Jurjevich, J. R. (2020). Contradictions of the climate-friendly city: New perspectives on eco-gentrification and housing justice. *International Journal of Urban and Regional Research, 44*(1), 145–165.

Shi, L., et al. (2016). Roadmap towards justice in urban climate adaptation research. *Nature Climate Change, 6*(2), 131–137.

Tol, R. S. (2005). Adaptation and mitigation: Trade-offs in substance and methods. *Environmental Science and Policy, 8*(6), 572–578.

Van der Heijden, J. (2018). From leaders to majority: A frontrunner paradox in built-environment climate governance experimentation. *Journal of Environmental Planning and Management, 61*(8), 1383–1401.

Van der Heijden, J. (2019). Studying urban climate governance: Where to begin, what to look for, and how to make a meaningful contribution to scholarship and practice. *Earth System Governance, 1*, 100005.

Wolfram, M., van der Heijden, J., Juhola, S., & Patterson, J. (2019). Learning in urban climate governance: Concepts, key issues and challenges. *Journal of Environmental Policy and Planning, 21*(1), 1–15.

Part II

Climate Urbanism and Transformative Action

6

Urban Climate Imaginaries and Climate Urbanism

Linda Westman and Vanesa Castán Broto

6.1 Introduction[1]

We live in a curious era in which optimism for cities as sites of transformative environmental and social action coexists with concerns about intensified urban social segregation, environmental degradation and a narrowing of urban policy around economic growth and protection. Climate urbanism has the potential to reinforce or challenge these trends, potentially opening up new forms of urban politics. Climate urbanism depends on the thoughts and actions of people who live in cities, and cities themselves are governed in relation to narratives about urban areas and their future. These narratives can help construct and justify interventions that respond to human concerns and shape cities.

[1] Note: this chapter is an abridged version of Westman, L. & Castán Broto, V. (under review) Urban Climate Imaginaries: The Framing of Cities in the Nationally Determined Contributions.

L. Westman (✉) • V. Castán Broto
Urban Institute, University of Sheffield, Sheffield, UK
e-mail: l.westman@sheffield.ac.uk; v.castanbroto@sheffield.ac.uk

V. Castán Broto et al. (eds.), *Climate Urbanism*,
https://doi.org/10.1007/978-3-030-53386-1_6

In this chapter, we argue that such narratives coalesce around specific urban climate imaginaries which are deployed to justify certain climate policies in urban environments. We explore the configuration of urban climate imaginaries in the international climate regime to examine the models of climate urbanism promoted under the umbrella of international climate policy. Our analysis shows that formal international climate policy documents centre around two dominant urban climate imaginaries: one relates to the concentration of socio-environmental concerns in urban areas and the need for central government control to address these; the other represents cities as agents of change and forerunners in climate action. Out of the two, the former dominates international climate policy discourses. This imaginary is linked to specific models of climate urbanism that become promoted in the context of the international climate regime. Specifically, this imaginary tends to support investment-intense agendas led by national governments, while it obscures a range of sectors and actors that might open alternative pathways of climate action in cities.

6.2 Urban Climate Imaginaries

Our definition of urban climate imaginaries draws, in part, on the concept of socio-technical imaginaries, which relates to powerful narrative devices oriented towards a normative vision of the future (Jasanoff and Kim 2009, 2015). Jasanoff and Kim (2009: 120) define socio-technical imaginaries as "collectively imagined forms of social life and social order reflected in the design and fulfilment of nation-specific scientific and/or technological projects." Socio-technical imaginaries are explicitly future-oriented, linked with but broader than public policy, related to action and performativity, and encoded in technologies and infrastructure (Jasanoff 2015). They capture how "science and technology become enmeshed in performing and producing diverse visions of the collective good" (Jasanoff 2015: 11). Socio-technical imaginaries are also embedded in the everyday use of technologies, as well as in the practices and norms associated with these (Strengers et al. 2019). Besides, our definition builds on the literature on urban imaginaries. When linked with formal policy, urban imaginaries can inspire specific forms of city-building strategies, such as visions based on mega-projects (Adama 2018) or demolition and regeneration (Mah 2012). In climate change policy, urban imaginaries have been related to normative visions of carbon-neutral cities, cities based in the new economy, or cities as a laboratory—all linked with different programs of investment and urban development (Tozer and Klenk 2018). Bloomfield (2006: 46) suggests that urban imaginaries focus on the sensory and emotional experience and practices, on the imprint of collective memory on how the city could be, on the different, often conflicting, social constructions of the city's future.

This definition diverts the notion of the collective away from government-expert blueprints for development, towards plural socio-political struggles for self-realization and recognition (Bloomfield 2006). Drawing on these two strands of thought, we define urban climate imaginaries as collective discourses surrounding the urban that reflect the aspirations of future visions, which also are embedded in institutions and socio-material practices. Imaginaries are created in narratives and reproduced in policy documents, but also enacted through other material forms (or other forms with material implications), for example, through the design of governance arrangements and associated investment strategies. As a result, the articulation of urban imaginaries in international narratives of climate change directly shapes opportunities to build just, resilient, and low-carbon cities, and influences what forms of climate urbanism we see emerging within the context of international climate policy.

There is a bidirectional relationship between situated experiences in cities and the formulation of urban imaginaries at the international level (Fig. 6.1). Perspectives and ideas of individuals in urban areas travel into global policy communities through their communication with higher-level authorities and their direct participation in international policy events (Cociña et al. 2019). Conversely, narratives and norms that circulate at the global level are transferred into national and local policy-making processes (Fisher 2014; McCann 2011). Figure 6.1 helps imagining these complex global-local relationships. However, these forms of influence are not linear. Instead, the concept of urban climate imaginaries provides an alternative to think of this relationship in complex ways, particularly in

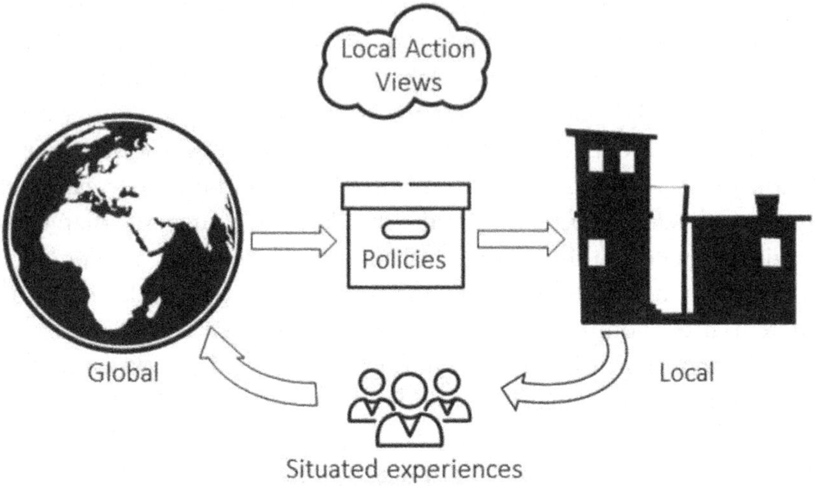

Fig. 6.1 Constitution of urban climate imaginaries and their materialization in narratives of action, policies, and situated experiences that shape both the international climate regime and the local context of action

terms of how representations of the city become attached to assemblages of ideas, actors, and material artefacts. These imaginaries are pervasive and last over time, through alignment with power structures, ways of articulating thought and speech, and entrenched ways of enacting policy among international decision-making communities.

6.3 The Urban Within the International Climate Regime

The IPCC Special Report on Global Warming of 1.5° highlighted the centrality of urban transformations as a means to increase climate action ambitions and deliver emission reductions beyond nationally committed targets (Masson-Delmotte et al. 2018). The 2015 Paris Agreement engages with urban areas, both in relation to increasing risks of climate change impacts in cities and opportunities for sustainable development through urbanization (Tollin and Hamhaber 2017). Policy initiatives, such as the inclusion of sub-national actors in the 2015 Paris Agreement, the NAZCA platform of sub-national action and the Durban group, identify cities as key actors involved in setting and implementing targets within the framework of the agreement (Hale 2016; Morgan and Northrop 2017).

In the context of international climate policy, the past decade has witnessed a fundamental shift from a focus on inter-state negotiations towards inclusion of a multiplicity of actors contributing to global emission reduction goals. The Paris Agreement was a watershed moment in this process, as it pivoted towards a voluntary system of contributions with the aim "to catalyse, support, and steer" activities of non-state actors towards transformative change (Chan et al. 2015: 465). Since then, the Nationally Determined Contributions (NDCs) constitute the mechanism through which each Party (nation-state) proposes targets for greenhouse gas emission reductions under the agreement (UNFCCC 2019). By 2019, 184 Parties had submitted NDCs to the UNFCCC, which are due to be updated in 2020. Here, cities represent one non-state actor expected to participate in mitigation and adaptation activities within national climate action agendas. At the same time, the NDCs have been criticized for poorly capturing the contributions of non-state actors to climate action agendas (Höhne et al. 2017). With regards to cities and urban areas, the Paris Agreement fails to provide details regarding any "specific responsibilities, capacities, governmental nature, or needs for support" (Satterthwaite et al. 2018: 2). While the formulation of the NDCs presented an opportunity for nation-states to elaborate their approach towards the sub-national and enhance participation of social organizations, this has yet to fully be realized. It is not yet clear how cities are mobilized in the process of formulating or enforcing climate action agendas in the context of the NDCs, or which views of the urban are being cemented within international climate policy debates. To unpack the range of views about desired interventions in urban environments in international climate policy, we analyse how cities are represented in the NDCs. The aim is to show how certain forms of climate urbanism are promoted through the entrenchment of urban climate imaginaries—and their links with specific forms of governance or investment strategies—in international climate policy debates.

Methodology

The analysis of urban climate imaginaries can be understood as a specific form of discourse analysis, which adopts text as the object of analysis, but assumes a wider connection between language use and modes of thought, action, social practices, and social relations. For this study, we compiled the NDCs from 184 Parties of the Paris Agreement (with the European Union NDC collectively representing its 28 member states) in the qualitative analysis software Nvivo. Our analysis followed an iterative process of coding, which built on three representations of the urban in international policy introduced by Parnell (2016) and others. These three representations included: (1) the urban as a site of sector-based concerns targeted for centralized intervention; (2) cities as a source of agency and leadership; and (3) urbanization as a phenomenon producing challenges and opportunities. Through in-depth exploration of these three representations and their links with policy and academic debates, we linked these representations with two urban climate imaginaries, as presented below.

Exploring Urban Imaginaries in the NDCs

Our analysis shows that the urban climate imaginaries deployed in the NDCs do not put cities and urban areas at the forefront of climate action. The dominant imaginary on display tends to restrict, rather than empower, actions in urban areas, and both imaginaries have implications for how we analyse and understand the emergence of a distinct form of climate urbanism.

The first imaginary, which is by far the stronger in the NDCs, presents urban areas as problematic sites of convergence of climate change-related concerns. This imaginary conflates representations of the urban as a site for sector-based issues and urbanization as a driver of socio-environmental degradation, encompassing most of the references to cities across all NDCs. This imaginary draws on fears of uncontrolled urban development and its associated consequences, including impacts of flooding, accumulation of waste, congested traffic, and deteriorating ecosystems. It appears to build on concerns with cities as an unstoppable force of consumption, which may connect with fears of over-population, resource depletion, and societal collapse. In doing so, this imaginary draws on ideas and images that have long circulated within international policy discourse, such as the slum eradication programs of the Millenium Development Goals or the extended history of a sector-based focus (such as on health and housing) in cities under colonial control (Parnell 2016). In some instances, the imaginary connects with ideals of progress, echoing ideas of cities as sanitized and hygienic (Brownell 2014; Swanson 1977), possibly drawing on supposedly universal visions of modernity (e.g. Kaika and Swyngedouw 2000). This imaginary serves to legitimize centralized intervention and control through reinforcement of sector-based national programs and governance interventions which concentrate powers within central governments. The emphasis is placed on how to generate local capacity through national programs, how to coordinate local action to deliver national outcomes, and how to streamline local planning processes to match central targets. While adopting the terminology of multi-level governance theory, this imaginary overlooks forms of horizontal collaboration based on mutual adjustment and authority-sharing. Instead, the approach follows concerns that uncoordinated, bottom-up responses will fail to deliver 'meaningful results' in the absence of powerful coordinating frameworks (Chan et al. 2015: 466; c.f. Tollin and Hamhaber 2017). Among other paradoxes produced by multi-level governance ideas (Westman et al. 2019), here we see how subverted versions of its terminology reinforce rather than disable top-down modes of governance. This imaginary designates of mitigation and adaptation concerns into specific sectorial urban agendas, and produces a synthetic, dichotomic divide between 'urban' and 'rural' climate change concerns. Mitigation programs in the transport and waste sectors are established as urban agendas, with housing or forestry much less often presented as urban. By contrast, rural areas are linked to poverty, food, and electrification. Both representations reflect the interests of aid programmes that imagine radically different sets of interventions in urban and rural areas. The NDCs appear to be reproducing imaginaries of aid communities, in which poverty becomes comparatively side-lined as an urban issue (Parnell 2016; Satterthwaite and Mitlin 2012). At the same time, waste and transport are sectors that have figured heavily in international financial programs, such as the clean development mechanism (CDM) (Sippel and Michaelowa 2010, 2013) and the Addis Ababa Declaration (UN General Assembly 2015). This highlights tactical reasons to emphasize these policy domains as urban concerns. This also shows the focus on managing emerging urban issues through 'infrastructuralization', that is, the transformation of resilience challenges through infrastructure provision. The form of climate urbanism that is supported through this dominant urban climate imaginary is thus characterized by the entrenchment of investment agendas in the transport and waste sectors, the concentration of authority within national government entities, and the marginalization of 'non-urban' domains (food and agriculture, forestry, as well as agendas linked with poverty). However, pre-figuring projects in this way prevents decision-makers from considering a broad range of interventions, leading, for example, to the deprioritization of forestry and agricultural projects from the urban sphere. The relative weakness of poverty-related issues in urban projects may also prevent engagement with issues of informality and marginalization.

Alongside the powerful imaginary of urban areas as sites that must be controlled, a few NDCs contain signs of an imaginary of cities as agents of change for sustainable futures. This imaginary signals the opportunity to address climate change in the future through innovations in urban form and function—through smart cities, low-emission zones, low-carbon construction, resilient buildings, circular economy, and green jobs. On the one hand, this imaginary may be associated with a host of support instruments and policy agendas long established within sustainable city discourses, alongside well-documented risks of neoliberal ideologies and exacerbated inequalities (Grossi and Pianezzi 2017; Marvin et al. 2015). On the other hand, this imaginary could be linked with opportunities to empower cities to realize this direction of change.

On the whole, however, the celebratory ideal of cities embodied by visionary Mayors taking action on a global arena void of ambition (e.g. Rosenzweig et al. 2010) is mainly absent in the NDCs. The agency of urban areas—as well as interests and abilities—is yet to be recognized. Likewise, the ability of cities to exercise climate leadership through transnational networks is almost wholly missing from the NDCs. One way to explain this is that imaginaries draw on pervasive ideals and images in international policy debates. As a result, popular ideas about cities as bold forerunners have not gained as much traction as consolidated perceptions of urban areas as locations of socio-environmental degradation. These results resonate with previous documentation of the reproduction of hegemonic and stereotypical representations of the 'urban' in global policy (Robin and Acuto 2018). Another explanation is that the imaginary that relates to sectorization and central control supports national and international political authority—therefore there are obvious strategic reasons to reproduce related narratives. In support of this perspective, there is a low saliency in the NDCs of the two causes that the local government and municipal authority constituency within the UNFCCC have been fighting for since their establishment. Regarding the first, fiscal arrangements, there is no evidence in the NDCs for plans to support urban areas through mechanisms to access streams of national or international funding. Regarding the second, political recognition in the international negotiations or devolved state power for municipal authorities, there is equally little or no interest. The dominant urban imaginary contained in these NDCs supports neither of these causes, as it appeals instead to an entrenchment of central power.

6.4 Conclusion

Urban climate imaginaries have implications for which forms of climate urbanism become known and established in the context of international policy. The literature on climate urbanism has documented a tendency towards the reproduction of investment-intense low-carbon infrastructure and agendas that seek to perpetuate economic growth, financialization, and privatization (Long and Rice 2019). These trends have been explained by the presence of neoliberal politics of the urban on a global level (Smith 2002), as well as the dynamics of interaction between political economies and politics of exclusion at the local level (Rice et al. 2020). The examination of urban climate imaginaries suggests that commonalities in urban climate action programs around the world are also shaped through the production and contestation of international policy discourses. Alongside coalitions of 'experts' in international organizations that promote certain visions of urbanism supported by specific forms of knowledge (Cociña et al. 2019; Rapoport 2015; Robin and Acuto 2018), imaginaries allow for ideas about the urban to persist over time and circulate across policy communities and scales of action. Urban imaginaries mobilized in the NDCs reveal a narrow interpretation of the future of cities, which favours influence of certain actors (central government) and prominence of certain sectors (waste, transport), yet preclude a range of interests and lived realities. One of our main concerns is that the consolidation of the current dominant imaginary in international climate policy will obscure and constrain a set of debates about the range of urban futures that are possible under conditions of climate change. This includes, for example, urban areas as sites to confront poverty, marginalization, and insecure housing, as well as issues related to agriculture, food production, and ecosystem protection. This may foreclose a broad range of socio-ecological imaginaries required to enable just urban transformations.

References

Adama, O. (2018). Urban imaginaries: Funding mega infrastructure projects in Lagos, Nigeria. *GeoJournal, 83*(2), 257–274.

Bloomfield, J. (2006). Researching the urban imaginary: Resisting the erasure of places. *Urban Mindscapes of Europe, 2*, 43–61.

Brownell, E. (2014). Seeing dirt in Dar es Salaam: Sanitation, waste, and citizenship in the postcolonial city. *The Arts of Citizenship in African Cities. Springer*, 209–229.

Chan, S., van Asselt, H., Hale, T., Abbott, K. W., Beisheim, M., Hoffmann, M., Guy, B., Höhne, N., Hsu, A., Pattberg, P., Pauw, P., Ramstein, C., & Widerberg, O. (2015). Reinvigorating international climate policy: A comprehensive framework for effective nonstate action. *Global Policy, 6*(4), 466–473.

Cociña, C., Frediani, A. A., Acuto, M., & Levy, C. (2019). Knowledge translation in global urban agendas: A history of research-practice encounters in the Habitat conferences. *World Development, 122*, 130–141.

Fisher, S. (2014). Exploring nascent climate policies in Indian cities: A role for policy mobilities? *International Journal of Urban Sustainable Development, 6*(2), 154–173.

Grossi, G., & Pianezzi, D. (2017). Smart cities: Utopia or neoliberal ideology?. *Cities, 69*, 79–85.

Hale, T. (2016). "All Hands on Deck": The Paris agreement and nonstate climate action. *Global Environmental Politics, 16*(3), 12–22.

Höhne, N., Kuramochi, T., Warnecke, C., Röser, F., Fekete, H., Hagemann, M., Day, T., Tewari, R., Kurdziel, M., & Sterl, S. (2017). The Paris agreement: Resolving the inconsistency between global goals and national contributions. *Climate Policy, 17*(1), 16–32.

Jasanoff, S. (2015). Future imperfect: Science, technology and the imaginations of modernity. In S. Jasanoff & S.-H. Kim (Eds.), *Dreamscapes of modernity: Sociotechnical imaginaries and the fabrication of power*. Chicago: University of Chicago Press.

Jasanoff, S., & Kim, S.-H. (2009). Containing the atom: Sociotechnical imaginaries and nuclear power in the United States and South Korea. *Minerva, 47*(2), 119.

Jasanoff, S., & Kim, S.-H. (2015). *Dreamscapes of modernity: Sociotechnical imaginaries and the fabrication of power*. Chicago: University of Chicago Press.

Kaika, M., & Swyngedouw, E. (2000). Fetishizing the modern city: The phantasmagoria of urban technological networks. *International Journal of Urban and Regional Research, 24*(1), 120–138.

Long, J., & Rice, J. L. (2019). From sustainable urbanism to climate urbanism. *Urban Studies, 56*(5), 992–1008.

Mah, A. (2012). Demolition for development: A critical analysis of official urban imaginaries in past and present UK cities. *Journal of Historical Sociology, 25*(1), 151–176.

Marvin, S., Luque-Ayala, A., & McFarlane, C. (Eds.). (2015). Smart urbanism: Utopian vision or false dawn?. Routledge.

Masson-Delmotte, V., Zhai, P., Pörtner, H.O., Roberts, D., Skea, J., Shukla, P.R., Pirani, A., Moufouma-Okia, W., Péan, C., Pidcock, R., Connors, S., Matthews, J.B.R., Chen, Y., Zhou, X., Gomis, M.I., Lonnoy, E., Maycock, T., Tignor, M., & Waterfield, T. (2018). Summary for Policymakers, IPCC.

McCann, E. (2011). Urban policy mobilities and global circuits of knowledge: Toward a research agenda. *Annals of the Association of American Geographers, 101*(1), 107–130.

Morgan, J., & Northrop, E. (2017). Will the Paris agreement accelerate the pace of change? *Wiley Interdisciplinary Reviews: Climate Change, 8*(5), e471.

Parnell, S. (2016). Defining a global urban development agenda. *World Development, 78*, 529–540.

Rapoport, E. (2015). Globalising sustainable urbanism: The role of international masterplanners. *Area, 47*(2), 110–115.

Rice, J. L. (2010). Climate, carbon, and territory: Greenhouse gas mitigation in Seattle, Washington. *Annals of the Association of American Geographers, 100*(4), 929–937.

Rice, J. L., Cohen, D. A., Long, J., & Jurjevich, J. R. (2020). Contradictions of the climate-friendly city: New perspectives on eco-gentrification and housing justice. *International Journal of Urban and Regional Research, 44*(1), 145–165.

Robin, E., & Acuto, M. (2018). Global urban policy and the geopolitics of urban data. *Political Geography, 66*, 76–87.

Rosenzweig, C., Solecki, W., Hammer, S. A., & Mehrotra, S. (2010). Cities lead the way in climate–change action. *Nature, 467*(7318), 909–911.

Satterthwaite, D., Dodman, D., Archer, A., & Brown, D. (2018). A look at how six global agreements support sustainable urban development. Retrieved from https://www.citiesalliance.org/newsroom/news/spotlight/look-how-six-global-agreements-support-sustainable-urban-development.

Satterthwaite, D., & Mitlin, D. (2012). *Urban poverty in the global south: Scale and nature*. London: Routledge.

Sippel, M., & Michaelowa, A. (2010). Does global climate policy promote low-carbon cities? Lessons learnt from the CDM. CIS Working Paper No. 4-2009. Available at SSRN https://ssrn.com/abstract=1551861 or https://doi.org/10.2139/ssrn.1551861.

Sippel, M., & Michaelowa, A. (2013). Financing a green urban economy: The potential of the Clean Development Mechanism (CDM). *The Economy of Green Cities, Springer*, 363–368.

Smith, N. (2002). New globalism, new urbanism: Gentrification as global urban strategy. *Antipode, 34*(3), 427–450.

Strengers, Y., Pink, S., & Nicholls, L. (2019). Smart energy futures and social practice imaginaries: Forecasting scenarios for pet care in Australian homes. *Energy research and Social Science, 48*, 108–115.

Swanson, M. W. (1977). The sanitation syndrome: Bubonic plague and urban native policy in the Cape Colony, 1900–1909. *The Journal of African History, 18*(3), 387–410.

Tollin, N., & Hamhaber, J. (2017). Sustainable Urbanization in the Paris Agreement. Comparative review for urban content in the Nationally Determined Contributions (NDCs), UNHABITAT, Kenya.

Tozer, L., & Klenk, N. (2018). Discourses of carbon neutrality and imaginaries of urban futures. *Energy Research and Social Science, 35*, 174–181.

UN General Assembly. (2015). Addis Ababa action agenda of the third international conference on financing for development, UN Development.

UNFCCC. (2019). Nationally Determined Contributions (NDCs). United Nations Framework Convention on Climate Change.

Westman, L. K., Broto, V. C., & Huang, P. (2019). Revisiting multi-level governance theory: Politics and innovation in the urban climate transition in Rizhao, China. *Political Geography, 70*, 14–23.

7

Institutional Dynamics of Transformative Climate Urbanism: Remaking Rules in Messy Contexts

James J. Patterson

7.1 Introduction

Transformations in urban governance are needed to address the multiple challenges of climate change. For example, this relates to decarbonization of systems of production and consumption (Bazaz et al. 2018; Bernstein and Hoffmann 2018; Bulkeley 2015); adaptation to climate impacts (e.g. floods, droughts, heat, fires) (Revi et al. 2014; World Bank 2011) and nonstationary boundary conditions (Craig 2010; Milly et al. 2008); the reshaping of infrastructure pathways that retrofit cities to avoid locking-in carbon intensive trajectories (Creutzig et al. 2016; NCE 2016); and the enhancement of equity and justice to ensure that climate action serves all members of society (Bulkeley et al. 2013; Shi et al. 2016). Institutions are pivotal to all these challenges. Advancing urban transformations entails making deliberate—and sometimes major—changes to institutions.

J. J. Patterson (✉)
Copernicus Institute of Sustainable Development, Utrecht University, Utrecht, Netherlands
e-mail: j.j.patterson@uu.nl

V. Castán Broto et al. (eds.), *Climate Urbanism*,
https://doi.org/10.1007/978-3-030-53386-1_7

However, institutions are often slow and resistant to change. This tension between institutional stability and revision comes into sharp focus in cities living with climate change. A key challenge of transformative climate urbanism is to understand how institutions can be deliberately changed within socially heterogeneous, historically encumbered, and politically contested settings.

Institutions are defined as "clusters of rights, rules and decision-making procedures that give rise to social practices, assign roles to the participants in these practices, and guide interactions among occupants of these roles" (Young et al. 2008: xxii), or "established and prevalent social rules that structure social interactions" (Hodgson 2006: 2). This involves both formal and informal aspects, which combine to produce the 'rules-in-use' (Ostrom 2005) in a given setting. Institutions are important for transformative climate urbanism because they condition decision-making and behaviour, create expectations, enshrine new precedents, and confer protections for vulnerable groups. But institutions are at most partially autonomous features of political life (March and Olsen 1983); they have casual influence on behaviour and social action, but at the same time, are themselves continuously re-shaped by political processes and patterns of behaviour (Pierson 2004).

The question addressed in this chapter is: *What types of institutional dynamics are likely to occur within urban climate transformations, and what are the implications for theorizing a transformative climate urbanism?* Taking inspiration from several bodies of literature including urban climate governance, environmental governance, transitions thinking, and political science, I develop a typology of institutional dynamics which are likely to coexist within processes of urban climate transformations, namely, Novelty, Uptake, Decline, Lock-in, Interplay, and Maintenance. This typology provides a novel set of analytical 'entry points' for studying the institutional dimensions of urban climate transformations. My departure point is a view of institutions as sites of political struggle where efforts to bring about deliberate changes inevitably encounter resistance, and produce friction with existing structures, arrangements, and practices. Agents within institutional settings continuously reinterpret and adapt institutions in the face of complex issues, at the same time as they affirm and enforce others (Mahoney and Thelen 2010). Sometimes agents may introduce radical new elements, or disrupt existing ones (Lawrence et al. 2009). This implies that struggles will inevitably ensue over urban climate transformations, with such struggles central to understanding the modalities of transformative climate urbanism.

7.2 Cities and Institutional Change

Cities are subject to dispersed authority located across multiple levels (e.g. municipal, state/province, national), generated through horizontal and transnational relationships (e.g. city networks), and held by private and civic actors. Institutions shape and channel the resulting complex interactions. Arguably, institutions of the state retain a central position in most cities, notwithstanding the many private, transnational (Bulkeley et al. 2014), and hybrid experimental (Bulkeley and Castán Broto 2013; Castán Broto and Bulkeley 2013; Hoffmann 2011) forms by which urban governance may also be accomplished. Nonetheless, even central actors such as municipal governments rarely have sole autonomy for climate action (Chap. 5). In this context, institutional change is likely to emerge in complex ways, involving interplay among agents, structures, and ideas.

Institutional change theories often emphasize the importance of change agents (Lawrence et al. 2009; Mahoney and Thelen 2010), and this is also widely observed empirically in urban climate governance literature (Anguelovski and Carmin 2011; Bassett and Shandas 2010; Carmin et al. 2012; Dilling et al. 2017; Roberts 2008). Some strands of institutional theory emphasize structural persistence through path dependency, whereby 'increasing returns' stabilize institutions due to accumulation of benefits and resources to invested actors, making reconfigurations unlikely without some sort of shock or disturbance (Pierson 2000). Ideational factors (e.g. discourse) may also cause changes through coordinative and communicative mechanisms among agents (Schmidt 2008, Chap. 6). Climate change shifts the material and even existential contexts for politics (Dryzek 2016), including urban governance. Practically, institutional changes may often occur through a variety of everyday means, such as legislators gritting through regulatory changes (Maltzman and

Shipan 2008), policymakers navigating windows of opportunity to secure policy changes (Kingdon 2014), bureaucrats persisting with policy implementation (Lipsky 1980), and communities cultivating new practices that enact and demonstrate climate action in urban settings (Castán Broto and Bulkeley 2013).

Yet, conceptually, our understanding of how deliberate institutional changes occur within urban climate transformations remains limited. As Dovers and Hezri (2010: 212) observed in regards to climate change adaptation: "institutions and institutional change are mentioned often but rarely specified", with "detailed discussion" particularly lacking at national and subnational scales. This gap persists a decade on. Three particular shortcomings exist in the urban climate literature. Firstly, institutions are often treated as a (static) independent variable to explain other social, political, and environmental changes, but rarely as a dependent variable (particularly institutional change). For example, institutions are invoked to explain success or failure of certain mitigation or adaptation actions, or to explain difficulties in advancing climate action more generally (e.g. Biagini et al. 2014; Carter et al. 2015; Gupta et al. 2010). Secondly, institutions are often treated as normative prescriptions concerning what should be implemented or changed to advance climate action. For example, it is common to see studies of climate change governance end with a recommendation that institutional change is needed, but give little indication of exactly what such changes might involve or how they can be realized (e.g. Barnett et al. 2015). Thirdly, institutions are sometimes treated as a consequence of innovative behaviour such as experimental social practices, whereby new social relations and/or norms become stabilized ('institutionalized') in a given setting (e.g. Anguelovski and Carmin 2011; Aylett 2015; Carmin et al. 2012). While this third aspect moves towards seeing institutional change as a dependent variable, it focuses on the incorporation of novelty into an existing setup, leaving open the question of how existing rules (and complex rule clusters) are deliberately changed.

7.3 Institutional Dynamics of Urban Climate Transformations

Various types of institutional dynamics are likely to emerge through deliberate efforts to bring about urban climate transformations. The term 'institutional dynamics' refers to fluidity in rules and rule clusters caused by attempts to introduce institutional changes, agential or structural resistance to such changes, as well as ongoing endogenous activity and adjustment. Table 7.1 summarizes a typology of coexisting institutional dynamics occurring within urban climate transformations. These are identified through a critical scoping review (following Grant and Booth 2009) across urban climate governance, sustainability transitions, and environmental governance literatures. The objective is to synthesize key thematic strands which often remain fragmented and posit an integrative

Table 7.1 Types of institutional dynamics in urban climate transformations

Type	Examples	Activity	Effect
Novelty	Innovation; Experimentation	Introduce new institutions (endogenous or exogenous sources)	Stimulate adoption of new institutions broadly
Uptake	Scaling; Diffusion; Catalysis	Adopt and propagate new institutions	Expand and entrench new institutions
Decline	Dismantling; Drift; Decay	Remove or destabilise existing institutions	Diminish influence of existing institutions
Lock-in	Path-dependency; Increasing returns	Resist new institutions due to systemic interdependencies	Failure to adopt new institutions
Interplay	Coherence; Conflict; Adjustment	Condition institutional activity by linkages with other domains	Contingency on other institutional arenas
Maintenance	Policy stability; Norm stability	Confer stability to existing institutions	Stability (productive or unproductive) persists

Adapted from Patterson (2020)

typology that ultimately enables hypothesis generation and testing of explanatory causal mechanisms. This typology helps to disaggregate the institutional dynamics of urban climate transformations and provides entry points for empirical analysis.

Dynamics of Novelty

Dynamics of novelty involve the introduction of new institutions as prototypes or exemplars, which can stimulate the widespread adoption of similar institutions, both within a city and in other cities elsewhere. This is one of the most richly explored areas of urban climate transformations, involving topics such as innovation and experimentation. Innovation literature focuses on institutional, technical, or social aspects, and often a combination of them. Institutional innovation typically centres on governmental activities (Patterson and Huitema 2019), for example, internal coordination of climate action (Aylett 2013; Bulkeley and Betsill 2013; Hughes 2017), and strategies for government coordination with the private sector and civil society (Burch 2010). Technical and social innovation are increasingly addressed jointly through approaches such as urban laboratories (Bulkeley et al. 2016; Voytenko et al. 2016) which focus on knowledge co-production between research and end users to support real-world uptake (Evans and Karvonen 2014). Experimentation literature is closely related, and highlights activities beyond the state, such as civil society, community, and business. It focuses on novel socio-technical practices (Bulkeley and Castán Broto 2013) which may lead to the development of novel governance arrangements (Hoffmann 2011), and afford permission to step outside the status quo (Farrelly and Brown 2011; Frantzeskaki et al. 2016). Dynamics of novelty have especially contributed to enlivening thinking in urban climate governance over the last decade, opening up avenues for studying and imagining urban climate transformations. Looking forward, further questions include: Under which institutional and political conditions do certain forms of novelty arise? By which institutional mechanisms do certain forms of novelty come into being? What are the politics of emphasizing novelty in urban climate transformations discourse, and who benefits or loses as a result?

Dynamics of Uptake

Dynamics of uptake involve the adoption and propagation of new institutions which expand their influence to become entrenched within the broader institutional fabric of urban governance. This is an area which has seen conceptual interest but much less empirical evidence in urban climate transformations (Chap. 5). Several concepts have been put forward to describe dynamics of uptake. For instance, scaling involves the expansion of a new institution (such as arising from an innovation or experiment) into broader existing institutional setups. This could include reshaping policies, laws, or decision-making procedures. In a study of urban sustainability in Asia, Bai et al. (2010) found that governmental linkages across levels were particularly important for scaling up experiments. More broadly, sustainability transitions literature argues that experiments may be scaled up to influence institutional landscapes and regimes (Geels 2002; Geels and Schot 2007). However, the mechanisms by which this scaling up occurs remain underdeveloped. The notion of diffusion refers to the spread of novelty via certain mechanisms of propagation (e.g. learning, emulation, competition, strategic dissemination) (Jordan and Huitema 2014; Shipan and Volden 2012; Strang and Soule 1998). This topic is explored in studies examining the spread of innovative practices through city networks (Hakelberg 2014; Lee and van de Meene 2012). Catalysis is a relatively less explored notion, which implies a degree of spontaneous movement or coordination arising in response to initial moves. For example, Bernstein and Hoffmann (2018: 191) suggest that subnational experiments may have catalytic political effects "by altering political dynamics within and across jurisdictions, markets, and/or carbon-intensive practices". Dynamics of uptake have been explored in different ways, using different metaphors and conceptual arguments. Nonetheless, they are all concerned with a similar core challenge of understanding the ways in which novelty is mobile and comes to have broader effects. Looking forward, much more attention needs to be paid to empirical studies of uptake to test and refine explanations which currently often remain highly conceptual. Key questions include: Under which institutional and political conditions do certain forms of uptake occur? What is the geography of uptake within and beyond cities' administrative boundaries (e.g. city-to-city learning, local-national government relations, relations between cities and international organizations)? By which institutional mechanisms does uptake occur? Are there specifically urban features of uptake which need to be considered in future theorizations?

Dynamics of Decline

Dynamics of decline involve the removal or destabilization of existing institutions leading to diminished reach and influence. This is an area which has so far seen little dedicated attention in urban climate transformations. Yet it is vital for understanding how disruptions create space for the uptake of new institutions, as well as the battles likely to play out over rapid shifts threatening the powers, resources, and positions of incumbents within particular institutional configurations (Brisbois 2019; Roberts et al. 2018). Sustainability transitions approaches, particularly the multi-level perspective, assume that disturbances in regimes occur either through niches putting pressure on a regime, or through disturbances emanating from the broader landscape (Geels and Schot 2007). However, this is unlikely to be sufficient, and transitions scholarship now increasingly calls for attention to studying the destabilization and decline of existing regimes (Geels 2014; Loorbach et al. 2017). Elsewhere, policy scholars identify policy dismantling as a topic requiring attention (Jordan et al. 2013). Historical institutional scholars place great focus on critical junctures as moments of path-breaking shifts (Pierson 2004), and long-standing policy change theory emphasizes shocks and windows of opportunity as moments of disruption and openness (Kingdon 2014: 1984). Institutional change theory identifies a mechanism of drift, where existing institutions become stagnant by not being adapted to changing external circumstances, rendering their impact diminished or dysfunctional within contexts that are increasingly mismatched with their design (Mahoney and Thelen 2010). Dynamics of decline have been little explored in urban climate transformations, and are an urgent research priority. There are many antecedents in broader political science and policy studies which can provide a starting point. Both conceptual and empirical work is required to develop thinking on dynamics of decline. Key questions include: Through which institutional mechanisms does decline occur in urban climate transformations? Under which institutional and political conditions is decline productive in the sense of allowing uptake of new institutions? To what extent and in which ways can insights about institutional and/or policy decline be applied in urban climate transitions?

Dynamics of Lock-in

Dynamics of lock-in involve resistance to new institutions due to systemic interdependencies which prevent the uptake of new institutions, resulting in their failure to be adopted. This is important to consider due to the multidimensional systems of production and consumption that needs to be altered by urban climate transformations. This means that deliberate (institutional) action may face resistance. One way this problem has been described is through the notion of 'carbon lock-in' (Unruh 2002, 2000). Seto et al. (2016) identify three core sources of carbon lock-in: (1) infrastructural and technological lock-in through investments in infrastructure with long lifespans and sunk costs, (2) institutional lock-in through path-dependency reinforced by incumbent interests, and (3) behavioural lock-in through individual and societal preferences, habits, and routines. Moreover, these three sources of lock-in are interconnected and mutually reinforcing. Bernstein and Hoffmann (2018: 189) argue that "diverse, decentralized responses" such as those of subnational actors can be crucial in disrupting carbon lock-in. Elsewhere, notions of path dependency are prominent in institutional literature, where processes of increasing returns accumulate benefits to those already benefiting, render the costs of switching paths relatively high, and mean that early departure points matter greatly down the line (Pierson 2004, 2000). The overall effect of lock-in is to produce a mismatch or clash between the existing system and new rules or practices introduced. Dynamics of lock-in have been moderately well theorized in climate change governance generally, although need to be explored specifically in relation to the emergence of a climate-oriented form of urbanism in cities. Looking forward, both conceptual and empirical work is required, particularly in order to find ways to 'unlock' lock-in (Bernstein and Hoffmann 2018). Key questions include: Through which institutional mechanisms does lock-in occur in urban climate transformations? Under which institutional and political conditions is lock-in stable/unstable? Given the complex and heterogeneous nature of cities, how might lock-in be 'unlocked'?

Dynamics of Interplay

Dynamics of interplay involve the conditioning of institutional activity in a particular area by linkages with other domains, which means that deliberate (institutional) action may be contingent on other institutional arenas. This is different from lock-in for two reasons: (1) interplay involves interaction with other lateral policy domains (e.g. health, environment, planning, social policy), and (2) interplay may not always be negative or constraining (e.g. co-benefits), yet still produce dynamics that need to be accounted for. Institutional interplay has been a topic of significant interest in environmental governance literature, particularly at the international level (Oberthür and Stokke 2011; van Asselt 2014; Young et al. 2008). It is inescapable in complex institutional settings, where the performance of any specific institution also depends on other connected institutions, and in the context of climate change which requires trans-sectorial responses. In environmental governance, a common distinction is made between vertical and horizontal interplay (Young et al. 2008), ideas which have also been applied in urban climate governance (Bulkeley and Betsill 2013). Some environmental governance scholars have further suggested mechanisms of interplay at an international scale, such as cognitive, commitment, behavioural, and impact interplay (Gehring and Oberthur 2008). Recent climate governance literature explores the notion of polycentricity, which reinterprets governance as an emergent phenomenon comprising multiple autonomous centres of decision-making and (possible) links between them (Jordan et al. 2018). This points towards forms of interplay such as competition, coordination, and mutual adjustment. Conversely, literature on fragmentation in environmental governance, which highlights incomplete patchworks of arrangements, points towards forms of interplay ranging from synergy to cooperation to conflict (Biermann et al. 2009). Interplay obviously remains a core issue in urban settings, where a wide range of policy and problem domains converge within a territorial and jurisdictional space (Chap. 5). Yet it requires substantial attention to advance this line of thinking, particularly to move beyond distinctions of horizontal/vertical interplay. One possibility could lie in rethinking urban regime theory (Mossberger and Stoker 2001; Rast 2015) by developing mid-range conceptual frameworks which are portable across cases (Ward 1996) (although this may not be appropriate beyond the Global North where this theory was developed). Alternatively, a more grounded approach could be taken to analyse interplay of institutional configurations within and across diverse urban contexts (particularly in the Global South), paying attention to emergent configurations as-they-exist, in order to develop new conceptualizations of interplay (e.g. such as building on the notion of 'messiness', following Castán Broto 2019).

Dynamics of Maintenance

Dynamics of maintenance involve the conferral of stability to existing institutions, particularly through replication by endogenous agents, which results in stability in potentially both productive or unproductive ways. This is different to lock-in for two reasons. Firstly, lock-in is generated by the wider system (e.g. technological, institutional, behavioural), whereas maintenance is conferred by endogenous actors based on conscious or unconscious behaviour to replicate institutions in everyday practice (Beunen and Patterson 2019). Thus, maintenance involves ideational factors, such as beliefs, worldviews, and culture (Patterson and Beunen 2019). Secondly, lock-in carries a negative connotation of being solely a barrier to deliberate action, whereas maintenance may also prevent change but it can also be positive in providing productive stability. For example, maintenance of foundational institutions such as democratic accountability and the rule of law underpin liberal democratic societies as a whole, and even a certain degree of policy and regulatory stability is usually considered vital for business investment including for low-carbon transitions (Beunen et al. 2017). A lack of maintenance of existing policy may lead to its undermining through regression or shifts in government agenda, as has been witnessed for biodiversity policy globally (Chapron et al. 2017) and some national climate policies (e.g. Crowley 2017). Overall, maintenance is clearly in inherent tension with other institutional dynamics (e.g. novelty, uptake) in the context of deliberate institutional change; in essence, the core tension between stability and flexibility (Beunen et al. 2017; Craig 2010). Dynamics of maintenance are not well understood in urban climate transformations, especially productive forms of maintenance seem to be almost entirely unstudied. Looking forward, this is an area which requires conceptualization. Key questions include: Through which institutional mechanisms does maintenance occur in urban climate transformations? Under which institutional and political conditions is maintenance productive or unproductive, and for whom? Which institutions should be maintained, and which need to be changed to advance urban climate transformations?

7.4 Conclusions

The typology outlined in this chapter brings together disparate lines of research that contribute to understanding the coexisting institutional dynamics involved in urban climate transformations. This provides opportunities for advancing research on the institutional dimensions of climate urbanism and its transformative potential, in order to understand its emergence in different locations—or lack thereof. By providing a set of entry points into the complex institutional dynamics of urban climate transformations, the typology reveals areas of tension both within certain categories (e.g. different conceptualizations of mobility within Uptake), and between categories (e.g. between Maintenance and Decline; Lock-in and Uptake). The chapter has several implications. It questions the common dichotomy between incremental and radical action: deliberate institutional interventions occur in complex settings, and trigger a variety of dynamics both for and against change. Not only can consequences of intervention often not be predicted, but sweeping reforms may often not be a panacea, or even a possibility. Often institutions take a long time to become socially and politically embedded. Furthermore, there is almost never a clean slate to begin with, meaning that interventions always contend with challenges posed by an existing setup with its own logic, history, and pattern of invested agents. Nonetheless, efforts towards bringing about deliberate change in the moments and the margins may be pursued within the everyday and often mundane realities of institutional life. Transformative change may often be underpinned by incremental changes which accumulate over time (Mahoney and Thelen 2010), potentially producing something radical but which cannot be fully foreseen at the outset (Patterson et al. 2017). Institutional change may be non-linear, and slow-moving causes may trigger thresholds leading to radical change under certain circumstances (Pierson 2004). Patterns of cumulation, coevolution, and non-linear outcomes are a key frontier for future research, which can be tackled through the typology presented here. Progressing urban climate transformations requires remaking the rules structuring urban governance. Without this, urban governance may remain largely unchanged, despite invocations for urgent transformation, and even the presence of diverse but often times ephemeral activity among various actors seeking to do things differently. The typology presented here helps to advance an agenda on climate urbanism by disaggregating the messy institutional dynamics occurring, stimulating debate about the institutional dimensions of transformation, and advancing novel ways in which these debates may be approached to help realize such transformations.

References

Anguelovski, I., & Carmin, J. (2011). Something borrowed, everything new: Innovation and institutionalization in urban climate governance. *Current Opinion in Environmental Sustainability, 3*(3), 169–175.

Aylett, A. (2013). The Socio-institutional Dynamics of Urban Climate Governance: A Comparative Analysis of Innovation and Change in Durban (KZN, South Africa) and Portland (OR, USA). *Urban Studies, 50*, 1386–1402.

Aylett, A. (2015). Institutionalizing the urban governance of climate change adaptation: Results of an international survey. *Urban Climate, 14*, 4–16.
Bai, X., Roberts, B., & Chen, J. (2010). Urban sustainability experiments in Asia: Patterns and pathways. *Environmental Science and Policy, 13*, 312–325.
Barnett, J., et al. (2015). From barriers to limits to climate change adaptation: Path dependency and the speed of change. Ecology and Society 20.
Bassett, E., & Shandas, V. (2010). Innovation and climate action planning: Perspectives from municipal plans. *Journal of the American Planning Association, 76*, 435–450.
Bazaz, A., et al. (2018). Summary for Urban Policymakers – What the IPCC Special Report on 1.5C Means for Cities. Indian Institute for Human Settlements.
Bernstein, S., & Hoffmann, M. (2018). The politics of decarbonization and the catalytic impact of subnational climate experiments. *Policy Sciences, 51*, 189–211.
Beunen, R., Patterson, J., & Van Assche, K. (2017). Governing for resilience: The role of institutional work. *Current Opinion in Environmental Sustainability, 28*, 10–16.
Beunen, R., & Patterson, J. J. (2019). Analysing institutional change in environmental governance: Exploring the concept of 'institutional work'. *Journal of Environmental Planning and Management, 62*, 12–29.
Biagini, B., Bierbaum, R., Stults, M., Dobardzic, S., & McNeeley, S. M. (2014). A typology of adaptation actions: A global look at climate adaptation actions financed through the Global Environment Facility. *Global Environmental Change, 25*, 97–108.
Biermann, F., Pattberg, P., Van Asselt, H., & Zelli, F. (2009). The fragmentation of global governance architectures: A framework for analysis. *Global environmental politics, 9*(4), 14–40.
Brisbois, M. C. (2019). Powershifts: A framework for assessing the growing impact of decentralized ownership of energy transitions on political decision-making. *Energy Research and Social Science, 50*, 151–161.
Bulkeley, H. (2013). *Cities and climate change*. Abingdon: Routledge.
Bulkeley, H. (2015). Can cities realise their climate potential? Reflections on COP21 Paris and beyond. *Local Environment, 20*, 1405–1409.
Bulkeley, H., & Betsill, M. (2013). Revisiting the urban politics of climate change. *Environmental Politics, 22*(1), 136–154.

Bulkeley, H., Carmin, J., Castán Broto, V., Edwards, G. A. S., & Fuller, S. (2013). Climate justice and global cities: Mapping the emerging discourses. *Global Environmental Change, 23*, 914–925.

Bulkeley, H., & Castán Broto, V. (2013). Government by experiment? Global cities and the governing of climate change. *Transactions of the Institute of British Geographers, 38*(3), 361–375.

Bulkeley, H., Andonova, L.B., Betsill, M.M., Compagnon, D., Hale, T., Hoffman, M.J., Newell, P., Paterson, M., Roger, C., & Vandeveer, S.D. (2014). Transnational climate change governance. Cambridge University Press, New York.

Bulkeley, H., Coenen, L., Frantzeskaki, N., Hartmann, C., Kronsell, A., Mai, L., Marvin, S., McCormick, K., van Steenbergen, F., & Palgan, Y. V. (2016). Urban living labs: Governing urban sustainability transitions. *Current Opinion in Environmental Sustainability, 22*, 13–17.

Burch, S. (2010). Transforming barriers into enablers of action on climate change: Insights from three municipal case studies in British Columbia, Canada. *Global Environmental Change, 20*, 287–297.

Carmin, J., Anguelovski, I., & Roberts, D. (2012). Urban Climate Adaptation in the Global South: Planning in an Emerging Policy Domain. *Journal of Planning Education and Research, 32*, 18–32. https://doi.org/10.1177/0739456X11430951.

Carter, J. G., Cavan, G., Connelly, A., Guy, S., Handley, J., & Kazmierczak, A. (2015). Climate change and the city: Building capacity for urban adaptation. *Progress in Planning, 95*, 1–66.

Castán Broto, V. (2019). *Urban energy landscapes*. Cambridge: Cambridge University Press.

Chapron, G., Epstein, Y., Trouwborst, A., & López-Bao, J. V. (2017). Bolster legal boundaries to stay within planetary boundaries. *Nature Ecology and Evolution, 1*, 0086.

Craig, R. K. (2010). "Stationarity is dead"- Long live transformation: Five principles for climate change adaptation law. *Harvard Environmental Law Review, 34*, 9–73.

Creutzig, F., Agoston, P., Minx, J. C., Canadell, J. G., Andrew, R. M., Le Quéré, C., Peters, G. P., Sharifi, A., Yamagata, Y., & Dhakal, S. (2016). Urban infrastructure choices structure climate solutions. *Nature Climate Change, 6*(12), 1054.

Crowley, K. (2017). Up and down with climate politics 2013–2016: The repeal of carbon pricing in Australia: Australia's climate politics 2013–2016. *Wiley Interdisciplinary Reviews: Climate Change, 8*, e458.

Dilling, L., Pizzi, E., Berggren, J., Ravikumar, A., & Andersson, K. (2017). Drivers of adaptation: Responses to weather- and climate-related hazards in 60 local governments in the Intermountain Western U.S. *Environment and Planning A, 49*, 2628–2648.

Dovers, S. R., & Hezri, A. A. (2010). Institutions and policy processes: The means to the ends of adaptation: Institutions and policy processes. *WIREs Climate Change, 1*, 212–231.

Dryzek, J. S. (2016). Institutions for the Anthropocene: Governance in a changing earth system. *British Journal of Political Science, 46*, 937–956.

Evans, J., & Karvonen, A. (2014). 'Give Me a Laboratory and I Will Lower Your Carbon Footprint!' - Urban Laboratories and the Governance of Low-Carbon Futures: Governance of low carbon futures in Manchester. *International Journal of Urban and Regional Research, 38*, 413–430.

Farrelly, M., & Brown, R. (2011). Rethinking urban water management: Experimentation as a way forward? *Global Environmental Change, 21*, 721–732.

Frantzeskaki, N., Kabisch, N., & McPhearson, T. (2016). Advancing urban environmental governance: Understanding theories, practices and processes shaping urban sustainability and resilience. *Environmental Science and Policy, 62*, 1–6.

Geels, F. W. (2002). Technological transitions as evolutionary reconfiguration processes: A multi-level perspective and a case-study. *Research policy, 31*, 1257–1274.

Geels, F. W. (2014). Regime resistance against low-carbon transitions: Introducing politics and power into the multi-level perspective. *Theory, Culture and Society, 31*, 21–40.

Geels, F. W., & Schot, J. (2007). Typology of sociotechnical transition pathways. *Research Policy, 36*, 399–417.

Gehring, T., & Oberthur, S. (2008). Interplay: Exploring institutional interaction. In O. R. Young, L. A. King, & H. Schroeder (Eds.), *Institutions and environmental change: Principle findings, applications, and research frontiers*. Cambridge, USA: The MIT Press.

Grant, M. J., & Booth, A. (2009). A typology of reviews: An analysis of 14 review types and associated methodologies: A typology of reviews, Maria J. Grant and Andrew Booth. *Health Information and Libraries Journal, 26*, 91–108.

Gupta, J., et al. (2010). The adaptive capacity wheel: A method to assess the inherent characteristics of institutions to enable the adaptive capacity of society. *Environmental Science and Policy, 13*, 459–471.

Hakelberg, L. (2014). Governance by diffusion: Transnational municipal networks and the spread of local climate strategies in Europe. *Global Environmental Politics, 14*, 107–129.

Hodgson, G. M. (2006). What are institutions? *Journal of Economic Issues, 40*, 1–25.

Hoffmann, M. (2011). *Climate governance at the crossroads experimenting with a global response after Kyoto*. New York: Oxford University Press.

Hughes, S. (2017). The politics of urban climate change policy: Toward a research agenda. *Urban Affairs Review, 53*(2), 362–380.

Jordan, A., Bauer, M. W., & Green-Pedersen, C. (2013). Policy dismantling. *Journal of European Public Policy, 20*, 795–805.

Jordan, A., & Huitema, D. (2014). Innovations in climate policy: The politics of invention, diffusion, and evaluation. *Environmental Politics, 23*, 715–734.

Jordan, A. J., Huitema, D., van Asselt, H., & Forster, J. (2018). *Governing climate change: Polycentricity in action?* Cambridge, MA: Cambridge University Press.

Kingdon, J. W. (2014). *Agendas, alternatives, and public policies* (2nd ed.). Boston, MA: Pearson new international edition. Pearson Education.

Lawrence, T. B., Suddaby, R., & Leca, B. (2009). *Institutional work: Actors and agency in institutional studies of organizations*. Cambridge: Cambridge University Press.

Lee, T., & van de Meene, S. (2012). Who teaches and who learns? Policy learning through the C40 cities climate network. *Policy Sciences, 45*, 199–220.

Lipsky, M., 1980. Street Level Bureaucracy: Dilemmas of the Individual in Public Services. Russell Sage Foundation.

Loorbach, D., Frantzeskaki, N., & Avelino, F. (2017). Sustainability transitions research: Transforming science and practice for societal change. *Annual Review of Environment and Resources, 42*, 20.

Mahoney, J., & Thelen, K. (2010). *Explaining Institutional Change: Ambiguity, Agency, and Power*. New York: Cambridge University Press.

Maltzman, F., & Shipan, C. R. (2008). Change, continuity, and the evolution of the law. *American Journal of Political Science, 52*, 252–267.

March, J. G., & Olsen, J. P. (1983). The New Institutionalism: Organizational Factors in Political Life. *American Political Science Review, 78*, 734–749.

Milly, P. C. D., et al. (2008). Stationarity is dead: Whither water management? *Science, 319*, 573–574.

Mossberger, K., & Stoker, G. (2001). The evolution of urban regime theory: The challenge of conceptualization. *Urban Affairs Review, 36*, 810–835.

NCE. (2016). *The Sustainable Infrastructure Imperative: Financing for Better Growth and Development (The 2016 New Climate Economy Report)*. London: The Global Commission on the economy and Climate, World Resources Institute, Washington DC, and Overseas Development Institute.

Oberthür, S., & Stokke, O. S. (Eds.). (2011). *Managing institutional complexity: Regime interplay and global environmental change, Institutional dimensions of global environmental change*. Cambridge, MA: MIT Press.

Ostrom, E. (2005). *Understanding institutional diversity*. Princeton, NJ: Princeton University Press.

Patterson, J., et al. (2017). Exploring the governance and politics of transformations towards sustainability. *Environmental Innovation and Societal Transitions, 24*, 1–16.

Patterson, J. J. (2020). *Remaking political institutions: Climate change and beyond. Elements in earth system governance*. Cambridge: Cambridge University Press.

Patterson, J. J., & Beunen, R. (2019). Institutional work in environmental governance. *Journal of Environmental Planning and Management, 11*, 1–11.

Patterson, J. J., & Huitema, D. (2019). Institutional innovation in urban governance: The case of climate change adaptation. *Journal of Environmental Planning and Management, 62*(3), 374–398.

Pierson, P. (2000). Increasing returns, path dependence, and the study of politics. *The American Political Science Review, 94*, 251.

Pierson, P. (2004). *Politics in time: History, institutions, and social analysis*. Princeton, NJ: Princeton University Press.

Rast, J. (2015). Urban regime theory and the problem of change. *Urban Affairs Review, 51*, 138–149.

Revi, A., et al. (2014). Urban areas (Chapter 8). In: Climate Change 2014: Impacts, adaptation, and vulnerability. Part A: Global and sectoral aspects. Contribution of Working Group II to the Fifth Assessment Report of the Intergovernmental Panel on Climate Change.

Roberts, C., Geels, F. W., Lockwood, M., Newell, P., Schmitz, H., Turnheim, B., & Jordan, A. (2018). The politics of accelerating low-carbon transitions: Towards a new research agenda. *Energy research & social science, 44*, 304–311.

Roberts, D. (2008). Thinking globally, acting locally — institutionalizing climate change at the local government level in Durban, South Africa. *Environment and Urbanization, 20*, 521–537.

Schmidt, V. A. (2008). Discursive institutionalism: The explanatory power of ideas and discourse. *Annual Review of Political Science, 11*, 303–326.

Seto, K. C., et al. (2016). Carbon Lock-In: Types, causes, and policy implications. *Annual Review of Environment and Resources, 41*, 425–452.

Shi, L., et al. (2016). Roadmap towards justice in urban climate adaptation research. *Nature Climate Change, 6*(2), 131–137.

Shipan, C. R., & Volden, C. (2012). Policy diffusion: Seven lessons for scholars and practitioners. *Public Administration Review, 72*, 788–796.

Strang, D., & Soule, S. A. (1998). Diffusion in organizations and social movements: From hybrid corn to poison pills. *Annual Review of Sociology, 24*, 265–290.

Unruh, G. C. (2000). Understanding carbon lock-in. *Energy Policy, 28*, 817–830.

Unruh, G. C. (2002). Escaping carbon lock-in. *Energy Policy, 30*, 317–325.

Van Asselt, H. (2014). The fragmentation of global climate governance: Consequences and management of regime interactions, New horizons in environmental and energy law. Edward Elgar, Cheltenham, UK; Northampton, MA, USA.

Voytenko, Y., McCormick, K., Evans, J., & Schliwa, G. (2016). Urban living labs for sustainability and low carbon cities in Europe: Towards a research agenda. *Journal of Cleaner Production, 123*, 45–54.

Ward, K. (1996). Rereading urban regime theory: A sympathetic critique. *Geoforum, 27*, 427–438.

World Bank. (2011). *Guide to climate change adaptation in cities.* Washington DC: World Bank.

Young, O. R., King, L. A., & Schroeder, H. (Eds.). (2008). *Institutions and environmental change: Principal findings, applications, and research frontiers.* Cambridge, MA: MIT Press.

8

Urban Resilience and the Politics of Development

Eric Chu

8.1 Introduction

The emergence of *climate urbanism* as a distinct planning project presents new tensions and challenges for municipal authorities around the world. In this chapter, I explore the phenomenon of *urban climate resilience* as an emblematic and pervasive case of this new planning logic. *Resilience* represents a planning ideal—one that embodies lofty visions of 'bouncing back' from and adapting to climate shocks while retaining economic prosperity and well-being—but offers only vague hints at its applicability in particular cities with diverse political, cultural, and human experiences (Meerow and Newell 2019; Vale 2014). In response, local rhetoric of and approaches to realizing climate resilience become enmeshed within the values and interests of political agents and forces that are constantly contesting the development trajectory of the city. For some, a logic of resilience leads city institutions, financing, as well as pathways of policy coordination, implementation, and evaluation to be

E. Chu (✉)
University of California—Davis, Davis, CA, USA
e-mail: ekch@ucdavis.edu

V. Castán Broto et al. (eds.), *Climate Urbanism*,
https://doi.org/10.1007/978-3-030-53386-1_8

filtered through the prism of proactive climate action (Bohland et al. 2019). However for others—especially those cities that are resource and capacity constrained—this planning logic raises new challenges as the political work of implementing resilience conflicts with established political coalitions, long-held development interests, and entrenched forms of socioeconomic exclusion (Bahadur and Tanner 2014b; Borie et al. 2019).

In this chapter, I draw upon the literature on climate resilience and adaptation to illustrate the real and conscious political decisions that cities make to design and implement risk management priorities. I offer vignettes from two Indian cities—Surat in Gujarat and Kochi in Kerala—whose municipal authorities have proactively pursued urban resilience as a core logic for planning and development over the past several years. Intergovernmental policy schemes within India have further incentivized the integration of resilience priorities with local industrial growth, social cohesion, infrastructure development, and smart city agendas. Through these vignettes, I highlight the rhetorical and symbolic power of resilience, and I also describe the degree to which its implementation can capture preexisting institutional structures and redirect development mandates in ways that exacerbate urban inefficiencies and inequalities. Resilience is therefore not only a classic case of *fast policy* (see McCann 2011; Peck & Theodore 2015), it also speaks to how contemporary modes of *climate urbanism* demand conscious and political decisions made within urban governance structures that may or may not correspond to real, equitable, and transformative climate action in cities. An exploration of these questions on how *climate urbanism* is instrumentalized may help to explain the process through which its ideals become reality on the ground in contested urban contexts.

8.2 Urban Resilience as a Contested Concept

In the 2000s, the United Nations, World Bank, and other multi/bilateral development aid organizations began to advocate for city-level actions to combat the growing climate challenge. This pivot was ostensibly to supplement ongoing efforts at the level of the nation-state, but which was increasingly impeded by disagreements between the world's largest greenhouse gas emitters and their unwillingness to offer financial and technological concessions to the rest. Investigations into the potential role of cities—championed by the World Banks Cities and Climate Change Programme, ICLEI-Local Governments for Sustainability, and other strategic transnational municipal networks (TMN) (Betsill and Bulkeley 2007)—yielded insights on the opportunities for locally driven emissions reduction, risk management, and inclusive decision-making (Bulkeley and Castán Broto 2013; Castán Broto 2017; Chu 2016a), and also highlighted the multiple layers of capacity and resource constraints at the local level (Anguelovski et al. 2014; Carmin et al. 2013).

Armed with these insights, a new arena of private and philanthropic actors set sight on disseminating and institutionalizing climate action in cities, especially across the global South, with programs such as the Rockefeller Foundation's Asian Cities Climate Change Resilience Network (ACCCRN) and subsequent 100 Resilient Cities (100RC) leading the charge (Chu 2018; Croese et al. 2020; Goh 2020; Moench et al. 2011). The idea was to mainstream climate actions into concurrent urban development initiatives, focusing on building cross-sectoral coalitions between housing, infrastructure, energy, health, and economic development functions while recasting these preexisting priorities as increasingly vulnerable given projected changes in urban temperature and precipitation levels (Chu 2016b; Sharma et al. 2014; Tanner et al. 2019). Recent studies show that this has led to governance innovations in cities, such as new institutions to support policy coordination, new internal and intergovernmental financing mechanisms, as well as innovative participatory decision-making arenas bringing together citizen and private sector interests (e.g. Anguelovski and Carmin 2011; Bulkeley et al. 2015; Castán Broto 2017; Chu et al. 2016; Hughes et al. 2018; Patterson and Huitema 2019). For Southern cities, the rise of climate-oriented transnational municipal networks (TMNs) served as conduits for the rapid dissemination of *resilience* as a generally accepted, scientifically backed, and privately financed priority, one that would offer TMN members and subscribers to this new planning logic high degrees of international notoriety, exposure to new sources of development capital, and an impression of radical, innovative decision-making based on a critical global challenge.

The rise of *resilience* in multilateral circles was emphatically embraced by local governments because of its applicability across numerous contexts (Borie et al. 2019). Its ideals and aspirations were undeniable, and its champions were able to propose a methodology to evaluate resilience-building outcomes and impacts—such as requiring cities to be more adaptable, flexible, and open to change (Cote and Nightingale 2011; Joseph 2013)—in relatively short project timeframes even in the absence of climate shocks and stressors (Brown et al. 2018). Resilience building, therefore, became an increasingly technical exercise, where capacities of individuals, communities, institutions, and businesses were enhanced through assessing exposure to climate impacts against socioeconomic risks and vulnerabilities to justify a pipeline of infrastructural interventions, financing mechanisms, and evaluative metrics for pilot projects. A neoliberal critique of the idea naturally emerged in response to the technocratic and managerial turn of the concept (Borie et al. 2019; Cretney 2014; Meerow and Newell 2019), especially in the context of widespread deconstruction of municipal powers, privatization of local services and infrastructure, and speculative forms of land management and development across the world (Anguelovski et al. 2016; Bahadur and Thornton 2015; Cretney 2014; Peck and Tickell 2002).

Critics of urban *resilience* were quick to note its ambiguous and inconsistent definition (Pizzo 2015), as well as how its initial framing can have significant bearing on its eventual outcomes (Béné et al. 2012; McEvoy et al. 2013). For example, Meerow and Stults (2016) showed that among diverse understandings of the concept across local governments in the United States, most practitioners favor an engineering-based 'bouncing back' definition of resilience. As a result, the application of the concept often lacked consideration of social and political power, agency, and inequality (Cretney 2014; Eakin et al. 2017; Friend and Moench 2015). Further, the concept has been applied through neoliberal governance ideologies that seek to decrease public authority, increase local self-reliance, and privatize social services to make cities more desirable in a global market economy (Cretney 2014; Godschalk 2003; Gunderson 2010; Joseph 2013; MacKinnon and Derickson 2013). Scholars are therefore increasingly documenting incidences of unequal resilience outcomes such as the creation of urban enclaves and exclusion zones shielded from climate risks through speculative infrastructure and displacement of lower-income communities (Anguelovski et al. 2019; Gould and Lewis 2018; Hodson and Marvin 2009; Keenan et al. 2018; Long and Rice 2019; Rice et al. 2020). Such interventions are often further justified through the non-recognition of marginalized communities' interests in planning processes or even the entrenchment of poverty for historically disadvantaged groups (Chu et al. 2016; Chu and Michael 2019; Ranganathan and Bratman 2019).

Despite these critiques, many governments often *do* realize the exclusionary tendencies of resilience in urban contexts, but are otherwise forced to seek trade-offs in resource and institutionally constrained environments while being dependent on externally derived development catalysts (Harris et al. 2017). Further, *resilience* as an urban planning logic has continued to increase in prevalence over the past several years. Prescriptive ideals such as resilient 'development' and resilient 'pathways' have taken root in policy spheres ranging from reports produced by the Intergovernmental Panel on Climate Change (e.g. IPCC 2018), the United Nation's New Urban Agenda (Birkmann et al. 2016), to various local and regional action plans. The concept remains powerful precisely because its definition is all-encompassing—that is, a sense of urban well-being and security in an uncertain world—and instrumentally, it is supported by a portfolio of planning tools and interventions that are systematically evaluated, scientifically backed, and informed by economic metrics (Bahadur and Thornton 2015; Weichselgartner and Kelman 2015). However, recent discussions on the merits of *resilience* are adamant that we pay greater attention to place, culture, justice, and identity (Borie et al. 2019). Advocates of more just and equitable forms of resilience note how planning processes must take into account differences in access to power, knowledge, and resources (Byskov et al. 2019; Matin et al. 2018; Meerow et al. 2019; Shi et al. 2016), as well as engage with underlying political economic structures to (re)frame more radical, transformative development agendas (Bahadur and Tanner 2014a; Ziervogel et al. 2017).

The following vignettes from two cities in India illustrate how they have navigated the challenges of implementing resilience—as both a planning project and theoretical concept—over the past decade. This history is best illustrated by the Rockefeller Foundation's foray into urban resilience programming through ACCCRN (2008–2014) and 100 Resilient Cities (100RC) (2013–2019). The Rockefeller Foundation worked with cities in India to design new political engagement approaches and strategies to mainstream climate priorities in existing infrastructure, service delivery, and municipal financial plans. Researchers have highlighted ACCCRN as a processual exemplar—in the form of 'shared learning dialogue' (ISET 2010)—which focused on bringing different stakeholders to the table to identify, discuss, and collaboratively design strategies for cities to manage emerging climate impacts and risks (Chu et al. 2016; Sharma et al. 2014). The vignettes show that urban resilience—as a particularly pervasive form of *climate urbanism*—indeed offers a new and attractive governance logic for constrained cities. However, the political and economic challenges of implementing resilience in reality cast long shadows when confronted with high levels of bureaucratic deficits, social inequality, and institutional inefficiencies.

8.3 Vignettes from Urban India

The tensions between applying economic versus environmental framings of resilience in Indian cities were apparent since the beginning. When the Rockefeller Foundation first approached Surat in 2008 to become a pilot city for their fledgling ACCCRN program, an argument was made that the city must rethink its development trajectory through the prism of climate change, especially since it had been susceptible to major flooding events in the past that caused significant economic damages and loss of life (Bhat et al. 2013; Chu 2018). Surat, on the west coast of India in the relatively wealthy state of Gujarat, realized that ensuring economic and social resilience priorities—especially in terms of protecting small- and medium-sized businesses and minimizing social conflict and strife during disaster events—were of utmost importance given the city's exponential population growth rates in recent decades. This rate was attributed to an influx of migrants attracted to the city's growing textiles, diamonds, and other small- and medium-sized industries. Economic and social resilience priorities ultimately framed ACCCRN's work with all pilot cities in

India. Through setting out a process of 'shared learning dialogue' (ISET 2010), ACCCRN-funded researchers consulted extensively with city stakeholders, ranging from community leaders to municipal decision-makers, to assess climate risks and discuss potential resilience-building strategies.

During the first several years of ACCCRN's intervention, the program focused on building coalitions and making sectoral policy leaders aware of changing flooding, sea level rise, and disease impacts on the city. The interactive 'shared learning dialogue' process led to a number of pilot projects on reinforcing flood protection infrastructure, designing housing blocks that employed new cooling technologies, and creating a mobile phone disease-tracking platform to safeguard public health during disaster events (Chu 2016b). The efforts culminated in the release of Surat's *City Resilience Strategy* in 2012. For the municipal government, the arrival of ACCCRN helped to create a common language of *resilience*, where line department leaders, planners, and engineers could come together and jointly discuss risks to city infrastructure and approaches to reducing the vulnerability of water pipelines, public buildings, and waste management facilities. The creation of this shared language was facilitated by ACCCRN and its local intermediaries, who informed the process with scientific models and maps that illustrated flood risks across time. Ultimately, this shared language of resilience morphed into a shared vision of what resilience in Surat *ought to be*. A technocratic approach to diagnosing risks and vulnerabilities in the urban system subsequently evolved into more value-based judgments on which parts of the city would be enhanced, protected, or rendered more (economically) resilient to climate impacts.

Protecting valuable infrastructure and capital assets from flood impacts was a key planning priority for Surat. However, the problem was that many low-income informal settlements were located along the floodplain—where land was marginal, cheap, and nearby to employment opportunities—so any decisions made on fortifying flood protection infrastructure would inevitably lead to the demolishing and relocation of these communities. For the city's elite industrial class, resilience meant protecting small- and medium-sized industries at its urban core from floods, so reinforcing a flood barrier to protect its primary economic

Fig. 8.1 Large-scale flood protection infrastructure along the Tapi River in Surat. (Photo taken by author)

engine took precedence over considerations of community displacement (Fig. 8.1). To be fair, the city also supported public housing and health awareness programming, and many of these efforts did indeed target middle- and low-income communities. However, the complaints coming from these communities were not only that they were not properly included in the decision-making chain, but that their interests, livelihoods, and neighbourhoods were disproportionately impacted by the city's flagship resilience interventions (Anguelovski et al. 2016; Chu and Michael 2019).

In the end, despite clear trade-offs in social equity, Surat's experience as an ACCCRN pilot city was widely heralded as a success case by its funders because climate resilience *was* ultimately embedded into the city's overall planning logic. The Rockefeller Foundation remarked on the city's robust process for integrating climate science into strategic planning as well as its approach to designing and prioritizing pilot projects that represented core economic interests. ACCCRN intervention also led to a tangible institutional legacy—in the form of the Surat Climate Change Trust—where a coalition of government and business leaders agreed to continue resilience programming in the city even after the conclusion of ACCCRN funding in 2014 (Chu 2016b). Municipal governments beyond the initial group of ACCCRN pilot cities began to take note of Surat's model of institutionalizing and implementing climate resilience amidst a wide array of difficult development challenges. ACCCRN attempted to scale-up efforts across India, but because its original approach in Surat was time and energy-intensive, these replication efforts were considered less successful given limited financing, time, and capacity support.

Kochi (formerly known as Cochin), in the southern state of Kerala, participated in this scaling-up effort (colloquially termed ACCCRN+). ICLEI's South Asia office was commissioned by the Rockefeller Foundation to conduct climate risk and vulnerability analyses across the city, and to identify key sectors and services that required attention. Kochi is a low-lying city susceptible to water-logging during monsoon seasons (Mathew et al. 2012). Kochi, together with the district of Ernakulam, serves as Kerala's economic and industrial hub. However, unlike Surat, much of Kochi's economic base relies on remittances from Persian Gulf countries, where there is a large Keralite diaspora who migrated in the 1970s and 1980s in search of economic opportunities in that region's burgeoning oil and natural gas industries (Zachariah et al. 2001).

In light of the city's increasing climate vulnerabilities, ACCCRN engaged with ICLEI-South Asia and Kochi's Centre for Heritage, Environment and Development (C-HED) to strengthen urban water supply, waste management, and stormwater drainage through building partnerships with the local municipal government, infrastructure managers, and community-based organizations such as welfare, self-help, and women's groups. The idea was to replicate ACCCRN's initial 'shared learning dialogue' approach to identify appropriate visions and interventions for the city. In this case, although climate vulnerabilities such as flooding and waterlogging were high on the government's planning agenda, the city cited a greater economic logic behind resilience building and advocated for establishing special economic zones in support of the city's shipping industry and port facilities. Like in Surat, a broader argument around economic power and investment took hold at the expense of equitable and inclusions visions of urban resilience. As one of the premier shipping ports in the country, Kochi's preference was to be able to manage flooding, coastal erosion, storm surges, and waterlogging to an extent that allowed for the normal functioning of critical industrial and trade infrastructure. The fact that the area also included critical wetland habitat and supported local fisher communities was of secondary concern.

Kochi's focus on economic resilience can be attributed to the municipal authority's need to re-evaluate the city's primary economic driver—in the form of transnational remittances—due to labour and industrial regulatory shifts in Gulf countries. Qatar, Kuwait, United Arab Emirates, and Saudi Arabia in particular have been enacting a series of legislations to reserve higher-paying jobs in the fossil fuels sector for their own citizens. For example, Saudi Arabia's *nitaqat* system (or colloquially know as *Saudization*) set forth a nationalization program which required the private sector to prioritize employment of native Saudi citizens (see Sadi 2013). The oil sector was targeted due to its traditional reliance on expatriate labor from South Asia, particularly more educated and skilled workers from Kerala. Gradually, as higher-paid managerial positions were reserved for native Saudis and lower-paid labor-intensive positions were taken up by migrants from Bangladesh, Indonesia, and other newcomers to the labor market, Keralites were forced to return to Kerala in droves. However, many returning expatriates, especially men, have spent many years living abroad and have reduced social and employment ties to their home country, and so become unemployed upon their return. In recent years, Kochi has become a primary centre for returning expatriate migrants, and without transnational remittances to fund its infrastructure and public services, the city has turned to other development strategies—such as (re)building shipping, small industrial, and information and communication services sectors—to generate revenue.

Figure 8.2 shows the Vallarpadam International Container Terminal, which is directly opposite to the old town of Fort Cochin, and was constructed to kick-start the city's container shipping and logics industry. The container terminal was built on a floodplain even though the city's resilience planning process identified the area as susceptible to sea level rise, storm surge, and estuarine flooding under current climate projections. The juxtaposition of traditional fishermen, municipal river transport, and the container terminal in the background is particularly jarring because it highlights the contradicting demands placed on low-lying,

Fig. 8.2 View from Fort Cochin towards Vypin and the Vallarpadam International Container Terminal. (Photo taken by author)

vulnerable pieces of land in Kochi. Finding a balance between economic and ecological resilience is difficult because the municipal government has prioritized strengthening the city's employment and industrial bases as opposed to long-term protection of coastal zones. From a policy standpoint, this decision seems practical due to immediate revenue demands to support the city's existing infrastructure, but also because the city is attempting to reassert its position in the global economy—in the form of a regional and global shipping hub—to replace its historic role as a node for transnational remittances of 'Gulf money'.

The vignettes from Surat and Kochi illustrate the conflicting dynamics and planning implications when cities interpret and reframe emerging resilience priorities within the context of their own development opportunities and constraints. Surat's resilience logic harnessed the entrepreneurial power of local industrial and economic elites to elevate the idea's legitimacy and practicability to catalyze broader political awareness. Kochi's resilience logic was more an indirect response to externally derived revenue shortfalls and a desire to more actively participate in the global economy. Efforts were made to ensure inclusion and equity, especially since both Surat and Kochi based their efforts on a decision-making model designed and promoted by the Rockefeller Foundation and its intermediaries, which include highly regarded global research institutions. Despite good intentions at integrating climate priorities into concurrent urban development priorities, *resilience*—as a distinct yet 'slippery' planning logic—was ultimately politically reconstituted in ways that diverged from its original ideals. Economic resilience, and to a lesser extent social resilience, retained its agenda-setting power within the municipality, and thus was applied to justify protective and productive infrastructure with exclusionary implications. Governance structures, including legislative tools and participatory arenas, were thus captured by this divergent entrepreneurial logic of urban resilience. Surat's experience with expensive flood protection infrastructure and Kochi's preference for constructing shipping container terminals (see Figs. 8.1 and 8.2), both primarily catering to private business and industrial priorities, are good examples of this.

8.4 Conclusion: The End of Urban 'Resilience'?

Surat eventually became a member city of the Rockefeller Foundation's 100 Resilient Cities (100RC) network, launched in 2013. 100RC supposedly built upon the lessons from ACCCRN but, in reality, shared little in common, not least in terms of staffing composition, programmatic incentives, and sources of finance. As the Rockefeller Foundation was eager to scale-up early ACCCRN successes into larger global impact through 100RC, representatives from Surat embarked on the international conference circuit to share experiences of the city's planning strategies, approaches to building coalitions, and tools for implementation. Surat's story of combating climate risks in a rapidly urbanizing, global

South, and dynamic setting lent power to the Rockefeller Foundation's own ideals, metrics, and *logic* of urban resilience. This logic was then backed by financial commitments to support Chief Resilience Officers within each 100RC member city, providing two years of salary for this high-level bureaucrat to raise awareness, build cross-sectoral connections, and lead resilience planning from within municipal governments (Bellinson and Chu 2019; Croese et al. 2020). Kochi was not yet a formal member of 100RC when the network itself was abruptly shut down in July 2019.

The main reasons cited for the termination of 100RC include a change in leadership atop the Rockefeller Foundation and the financial implications of operating the network and supporting staff and programs within member cities. The issue of leadership change is of note in this case since this highlights the temporary nature of *fast policies* that rely on particular agents to continuously perpetuate and justify the value of these ideas. However, in the case of urban resilience, the idea already had sufficient political traction around the world, with the Rockefeller Foundation ultimately becoming only one of many philanthropic and private funders in the arena. Although some 100RC member cities were surprised by the sudden closure of the network—particularly those cities who were still benefiting from direct salary contributions—the resilience agenda has already taken root globally, where ideals such as 'bouncing-back', economic prosperity, risk management, and science-informed policy-making in the context of climate change were sufficiently embedded within existing planning and development portfolios in cities.

The intention behind this chapter is not to directly critique the role of the Rockefeller Foundation as an agent of resilience thinking or to question whether ACCCRN or 100RC networks somehow directly promoted exclusionary urban development in Surat, Kochi, and other member cities. This chapter is simply trying to illustrate how *resilience*—as a distinct and fast-moving planning logic that is increasingly prevalent around the world—embodies a particular set of language, ideals, and operational instruments that are easily captured and redirected, for better or worse, given its slippery definition, perceived universality, rhetorical power, and advocacy by well-funded patrons. The *political work* behind climate resilience, and by extension *climate urbanism*, entails its enmeshment within the values and interests of political agents and forces that are constantly contesting the development trajectory of the city. Taking the Rockefeller Foundation as an example, its various programs did manage to coalesce ideas, knowledge, and local buy-in around important issues such as economic well-being, social cohesion, environmental protection (Brown 2018). It also succeeded in building cross-sectoral coalitions within cities to translate planning ideals and climate science into on-the-ground action. Surat benefited from being a global exemplar, which catalyzed its leadership role in subsequent domestic smart cities, urban health resilience, and housing provision schemes.

The story, however, is not all rosy. Resilience, as a manifestation of *climate urbanism*, is subjected to the same critiques of exclusion and inequity as other large-scale infrastructure and technocratic planning projects. The way in which this new planning logic is embedded within cities can conflict with established political coalitions, long-held development interests, and entrenched forms of socioeconomic exclusion. When situated in highly contested and unequal contexts in the global South, urban resilience often places economy, entrepreneurialism, and self-interest as the primary motivators instead of ecological well-being and climate protection. Others note how its vague definition gives it speculative attributes that invite privatization, heralded by a global technocratic elite and a distinctly colonial North-South orientation to technological and financial networks (Chu 2018; Cretney 2014; Goh 2020). Despite all of this, the vignettes show that cities are in fact mostly aware of these trade-offs and challenges of resilience thinking (see Chelleri et al. 2015; Harris et al. 2017), but are structurally disincentivized to rectify its negative consequences because doing so runs contrary to the mantras of good governance, social stability, and economic prosperity.

The politics of urban resilience therefore refers to a process of negotiating and navigating in between its trade-offs, through time, and against contending entrenched ideological priorities and interests that seek to capture its benefits. Those who use the abrupt shutdown of 100RC as evidence of 'the end of urban resilience' may actually be discounting the practical realities and rhetorical power that resilience offers to cities that lack capacity, resources, or even a clear future vision for what the city *ought to be*. So, the issue in question is not really whether resilience is a legitimate framing concept or not, but rather under what conditions can resilience be designed and implemented in ways that do not reify the speculative, exclusionary, and entrepreneurial tendencies of urban governance around the world. Future research on *climate urbanism* must, therefore, explore approaches to reframing technocratic urban resilience into a development logic with reflexive, inclusive, redistributive, and emancipatory values.

References

Anguelovski, I., & Carmin, J. (2011). Something borrowed, everything new: Innovation and institutionalization in urban climate governance. *Current Opinion in Environmental Sustainability, 3*(3), 169–175.

Anguelovski, I., Chu, E., & Carmin, J. (2014). Variations in approaches to urban climate adaptation: Experiences and experimentation from the global South. *Global Environmental Change, 27*, 156–167.

Anguelovski, I., Shi, L., Chu, E., Gallagher, D., Goh, K., Lamb, Z., et al. (2016). Equity impacts of urban land use planning for climate adaptation: Critical perspectives from the global north and south. *Journal of Planning Education and Research, 36*(3), 333–348.

Anguelovski, I., Connolly, J. J. T., Pearsall, H., Shokry, G., Checker, M., Maantay, J., et al. (2019). Opinion: Why green "climate gentrification" threatens poor and vulnerable populations. *Proceedings of the National Academy of Sciences, 116*(52), 26139–26143.

Bahadur, A., & Tanner, T. (2014a). Transformational resilience thinking: Putting people, power and politics at the heart of urban climate resilience. *Environment and Urbanization, 26*(1), 200–214.

Bahadur, A. V., & Tanner, T. (2014b). Policy climates and climate policies: Analysing the politics of building urban climate change resilience. *Urban Climate, 7*, 20–32.

Bahadur, A. V., & Thornton, H. (2015). Analysing urban resilience: A reality check for a fledgling canon. *International Journal of Urban Sustainable Development, 7*(2), 196–212.

Bellinson, R., & Chu, E. (2019). Learning pathways and the governance of innovations in urban climate change resilience and adaptation. *Journal of Environmental Policy and Planning, 21*(1), 76–89.

Béné, C., Wood, R. G., Newsham, A., & Davies, M. (2012). Resilience: New Utopia or New Tyranny? Reflection about the potentials and limits of the concept of resilience in relation to vulnerability reduction programmes. *IDS Working Papers, 2012*(405), 1–61.

Betsill, M. M., & Bulkeley, H. (2007). Looking back and thinking ahead: A decade of cities and climate change research. *Local Environment, 12*(5), 447–456.

Bhat, G. K., Karanth, A., Dashora, L., & Rajasekar, U. (2013). Addressing flooding in the city of Surat beyond its boundaries. *Environment and Urbanization, 25*(2), 429–441.

Birkmann, J., Welle, T., Solecki, W., Lwasa, S., & Garschagen, M. (2016). Boost resilience of small and mid-sized cities. *Nature, 537*(7622), 605–608.

Bohland, J., Davoudi, S., & Lawrence, J. L. (Eds.). (2019). *The resilience machine*. New York and London: Routledge.

Borie, M., Pelling, M., Ziervogel, G., & Hyams, K. (2019). Mapping narratives of urban resilience in the global south. *Global Environmental Change, 54*, 203–213.

Brown, A. (2018). Visionaries, translators, and navigators: Facilitating institutions as critical enablers of urban climate change resilience. In S. Hughes, E. K. Chu, & S. G. Mason (Eds.), *Climate change in cities: Innovations in multi-level governance* (pp. 229–253). Cham, Switzerland: Springer International Publishing.

Brown, C., Shaker, R. R., & Das, R. (2018). A review of approaches for monitoring and evaluation of urban climate resilience initiatives. *Environment, Development and Sustainability, 20*(1), 23–40.

Bulkeley, H., & Castán Broto, V. (2013). Government by experiment? Global cities and the governing of climate change. *Transactions of the Institute of British Geographers, 38*(3), 361–375.

Bulkeley, H., Castán Broto, V., & Edwards, G. A. S. (2015). *An urban politics of climate change: Experimentation and the governing of socio-technical transitions*. New York and London: Routledge.

Byskov, M. F., Hyams, K., Satyal, P., Anguelovski, I., Benjamin, L., Blackburn, S., et al. (2019). An agenda for ethics and justice in adaptation to climate change. *Climate and Development*, 1–9. https://doi.org/10.1080/17565529.2019.1700774

Carmin, J., Dodman, D., & Chu, E. (2013). *Urban climate adaptation and leadership: From conceptual understanding to practical action (No. 2013/26)*. Paris: Organisation for Economic Co-operation and Development (OECD).

Castán Broto, V. (2017). Urban governance and the politics of climate change. *World Development, 93*, 1–15.
Chelleri, L., Waters, J. J., Olazabal, M., & Minucci, G. (2015). Resilience trade-offs: Addressing multiple scales and temporal aspects of urban resilience. *Environment and Urbanization, 27*(1), 181–198.
Chu, E. (2016a). The governance of climate change adaptation through urban policy experiments. *Environmental Policy and Governance, 26*(6), 439–451.
Chu, E. (2016b). The political economy of urban climate adaptation and development planning in Surat, India. *Environment and Planning C: Government and Policy, 34*(2), 281–298.
Chu, E. (2018). Transnational support for urban climate adaptation: Emerging forms of agency and dependency. *Global Environmental Politics, 18*(3), 25–46.
Chu, E., Anguelovski, I., & Carmin, J. (2016). Inclusive approaches to urban climate adaptation planning and implementation in the Global South. *Climate Policy, 16*(3), 372–392.
Chu, E., & Michael, K. (2019). Recognition in urban climate justice: Marginality and exclusion of migrants in Indian cities. *Environment and Urbanization, 31*(1), 139–156.
Cote, M., & Nightingale, A. J. (2011). Resilience thinking meets social theory: Situating social change in socio-ecological systems (SES) research. *Progress in Human Geography, 36*(4), 475–489.
Cretney, R. (2014). Resilience for whom? Emerging critical geographies of socio-ecological resilience. *Geography Compass, 8*(9), 627–640.
Croese, S., Green, C., & Morgan, G. (2020). Localizing the sustainable development goals through the lens of urban resilience: Lessons and learnings from 100 resilient cities and Cape Town. *Sustainability, 12*(2), 550.
Eakin, H., Bojórquez-Tapia, L. A., Janssen, M. A., Georgescu, M., Manuel-Navarrete, D., Vivoni, E. R., et al. (2017). Opinion: Urban resilience efforts must consider social and political forces. *Proceedings of the National Academy of Sciences, 114*(2), 186–189.
Friend, R., & Moench, M. (2015). Rights to urban climate resilience: Moving beyond poverty and vulnerability. *Wiley Interdisciplinary Reviews: Climate Change, 6*(6), 23.
Godschalk, D. R. (2003). Urban hazard mitigation: Creating resilient cities. *Natural Hazards Review, 4*(3), 136–143.
Goh, K. (2020). Flows in formation: The global-urban networks of climate change adaptation. *Urban Studies, 57*(11), 2222–2240.

Gould, K. A., & Lewis, T. L. (2018). From green gentrification to resilience gentrification: An example from Brooklyn. *City and Community, 17*(1), 12–15.
Gunderson, L. (2010). Ecological and human community resilience in response to natural disasters. *Ecology and Society, 15*(2).
Harris, L. M., Chu, E. K., & Ziervogel, G. (2017). Negotiated resilience. *Resilience, 6*(3), 1–19.
Hodson, M., & Marvin, S. (2009). "Urban ecological security": A new urban paradigm? *International Journal of Urban and Regional Research, 33*(1), 193–215.
Hughes, S., Chu, E. K., & Mason, S. G. (Eds.). (2018). *Climate change in cities: Innovations in multi-level governance.* Cham: Springer International Publishing.
IPCC. (2018). *Global Warming of 1.5 °C. An IPCC Special Report on the impacts of global warming of 1.5 °C above pre-industrial levels and related global greenhouse gas emission pathways, in the context of strengthening the global response to the threat of climate change, sustainable development, and efforts to eradicate poverty* [Masson-Delmotte, V., P. Zhai, H.-O. Pörtner, D. Roberts, J. Skea, P.R. Shukla, A. Pirani, W. Moufouma-Okia, C. Péan, R. Pidcock, S. Connors, J.B.R. Matthews, Y. Chen, X. Zhou, M.I. Gomis, E. Lonnoy, T. Maycock, M. Tignor, and T. Waterfield (Eds.)]. Geneva, Switzerland: Intergovernmental Panel on Climate Change (IPCC).
ISET. (2010). *The shared learning dialogue: Building stakeholder capacity and engagement for climate resilience action.* Boulder, CO: Institute for Social and Environmental Transition.
Joseph, J. (2013). Resilience as embedded neoliberalism: A governmentality approach. *Resilience, 1*(1), 38–52.
Keenan, J. M., Hill, T., & Gumber, A. (2018). Climate gentrification: From theory to empiricism in Miami-Dade County, Florida. *Environmental Research Letters, 13*(5).
Long, J., & Rice, J. L. (2019). From sustainable urbanism to climate urbanism. *Urban Studies, 56*(5), 992–1008.
MacKinnon, D., & Derickson, K. D. (2013). From resilience to resourcefulness. *Progress in Human Geography, 37*(2), 253–270.
Mathew, S., Trück, S., & Henderson-Sellers, A. (2012). Kochi, India case study of climate adaptation to floods: Ranking local government investment options. *Global Environmental Change, 22*(1), 308–319.
Matin, N., Forrester, J., & Ensor, J. (2018). What is equitable resilience? *World Development, 109*, 197–205.

McCann, E. (2011). Urban policy mobilities and global circuits of knowledge: Toward a research agenda. *Annals of the Association of American Geographers, 101*(1), 107–130.

McEvoy, D., Fünfgeld, H., & Bosomworth, K. (2013). Resilience and climate change adaptation: The importance of framing. *Planning Practice and Research, 28*(3), 280–293.

Meerow, S., & Newell, J. P. (2019). Urban resilience for whom, what, when, where, and why? *Urban Geography, 40*(3), 309–329.

Meerow, S., Pajouhesh, P., & Miller, T. R. (2019). Social equity in urban resilience planning. *Local Environment, 24*(9), 793–808.

Meerow, S., & Stults, M. (2016). Comparing conceptualizations of urban climate resilience in theory and practice. *Sustainability, 8*(7), 701.

Moench, M., Tyler, S., & Lage, J. (Eds.). (2011). *Catalyzing urban climate resilience: Applying resilience concepts to planning practice in the ACCCRN program 2009–2011.* Boulder, CO: Institute for Social and Environmental Transition (ISET).

Patterson, J. J., & Huitema, D. (2019). Institutional innovation in urban governance: The case of climate change adaptation. *Journal of Environmental Planning and Management, 62*(3), 374–398.

Peck, J., & Theodore, N. (2015). *Fast policy: Experimental statecraft at the thresholds of neoliberalism.* Minneapolis: University Of Minnesota Press.

Peck, J., & Tickell, A. (2002). Neoliberalizing space. *Antipode, 34*(3), 380–404.

Pizzo, B. (2015). Problematizing resilience: Implications for planning theory and practice. *Cities, 43*, 133–140.

Ranganathan, M., & Bratman, E. (2019). From urban resilience to abolitionist climate justice in Washington, DC. *Antipode,* (early view). https://doi.org/10.1111/anti.12555.

Rice, J. L., Cohen, D. A., Long, J., & Jurjevich, J. R. (2020). Contradictions of the climate-friendly city: New perspectives on eco-gentrification and housing justice. *International Journal of Urban and Regional Research, 44*(1), 145–165.

Sadi, M. A. (2013). The implementation process of nationalization of workforce in Saudi Arabian private sector: A review of "Nitaqat Scheme.". *American Journal of Business and Management, 2*(1), 37.

Sharma, D., Singh, R., & Singh, R. (2014). Building urban climate resilience: Learning from the ACCCRN experience in India. *International Journal of Urban Sustainable Development, 6*(2), 133–153.

Shi, L., Chu, E., Anguelovski, I., Aylett, A., Debats, J., Goh, K., et al. (2016). Roadmap towards justice in urban climate adaptation research. *Nature Climate Change, 6*(2), 131–137.

Tanner, T., Zaman, R. U., Acharya, S., Gogoi, E., & Bahadur, A. (2019). Influencing resilience: The role of policy entrepreneurs in mainstreaming climate adaptation. *Disasters, 43*(S3), S388–S411.

Vale, L. J. (2014). The politics of resilient cities: Whose resilience and whose city? *Building Research and Information, 42*(2), 191–201.

Weichselgartner, J., & Kelman, I. (2015). Geographies of resilience: Challenges and opportunities of a descriptive concept. *Progress in Human Geography, 39*(3), 249–267.

Zachariah, K. C., Mathew, E. T., & Rajan, S. I. (2001). Social, economic and demographic consequences of migration on Kerala. *International Migration, 39*(2), 43–71.

Ziervogel, G., Pelling, M., Cartwright, A., Chu, E., Deshpande, T., Harris, L., et al. (2017). Inserting rights and justice into urban resilience: A focus on everyday risk. *Environment and Urbanization, 29*(1), 123–138.

9

Two Cheers for "Entrepreneurial Climate Urbanism" in the Conservative City

Corina McKendry

9.1 Introduction

The New Climate Urbanism is reshaping what cities are and can be in the face of a changing climate. Other chapters in this volume illustrate how, as climate change gains a predominant place on urban agendas, the changes this engenders may be transformative or exclusionary, creating spaces for climate justice or elite, green enclaves that exclude many. Understanding the contexts under which these different responses to climate change emerge is one of the promises of the climate urbanism research agenda. This chapter contributes to this task through a slightly different approach, by investigating a city that does not have a political commitment to addressing climate change yet nevertheless is taking important steps to reduce its carbon emissions. In doing so, I strive to illuminate the potential importance of pro-growth urban development

C. McKendry (✉)
Colorado College, Colorado Springs, CO, USA
e-mail: cmckendry@coloradocollege.edu

V. Castán Broto et al. (eds.), *Climate Urbanism*,
https://doi.org/10.1007/978-3-030-53386-1_9

strategies for climate urbanism, particularly in conservative places where the issue of climate change has little political traction.

Colorado Springs, with a population of nearly half a million residents, has no climate change mitigation plan, nor has it adopted a municipal sustainability plan of which greenhouse gas reduction is a part. Numerous past and present city council members (who also are the board of directors of the municipally owned utility) have been outspoken climate deniers. In 2017, when mayors across the United States were criticizing President Trump's plan to remove the US from the Paris Climate Agreement, the mayor of Colorado Springs dismissed these actions as little more than political posturing, and climate change an inappropriate issue for mayors to take on (Zubeck 2017). Yet in the past few years, and after significant community effort, some changes are being made in Colorado Springs that are slowly reshaping the built environment of the city in ways that will reduce carbon emissions from electricity and transportation. These include downtown infill development after decades of urban sprawl, the scheduled closure of one of its two coal-fired power plants, and a growing number of dedicated bike lanes on city center streets.

There is nothing unique or even remotely bold about any of these changes—in many ways Colorado Springs is just starting to adopt the "smart growth" ideas of two decades ago (Daniels 2001). Yet they are the result of years of local effort in a political context in which these modest steps toward urban sustainability have been deeply controversial. Reflecting on theorizations and critiques of climate urbanism and, more broadly, of urban greening as an economic development strategy, I ask whether in some circumstances the growth imperative that is the foundation of these approaches to city sustainability is necessary. In cities like Colorado Springs where there is no local government support for climate action (radical or otherwise), lower-carbon development as a growth strategy is an important political victory that may lay the groundwork for moving toward a still-being-imagined transformative climate urbanism. As such, the limitations and potential of climate urbanism must be contextualized within the actually existing politics of what it takes to spur change in particular places.

9.2 Pro-growth City Environmentalism

One of the notable shifts that has occurred with the rise of liberal environmentalism over the past three decades is the belief that environmental protection and economic growth can be compatible (Bernstein 2001). City officials, who in the United States are heavily dependent on local tax revenue, have embraced this idea of green growth with particular enthusiasm (Fitzgerald 2010; Hammer et al. 2011; Bloomberg and Pope 2017). Yet, a number of critiques have emerged of the various growth-oriented forms of city environmentalism. One is that these efforts selectively incorporate urban greening and sustainability into a neoliberal development model as a form of green urban entrepreneurialism to attract young professionals, business headquarters, and tourists. This has, at best, modest environmental benefits and, more likely, does nothing to address the ecological and social contradictions of twenty-first-century capitalism (While et al. 2004; Gibbs and Krueger 2007; Jonas and While 2007; Jonas et al. 2011; Janos and McKendry 2014). A second line of critique is that the technological fixes that urban sustainability, including climate urbanism, offers depoliticize the issue and fails to recognize that both climate change and our responses to it are deeply political and connected to broader questions of power and inequalities. Depoliticized technological fixes and arguments that "we are all in this together" are therefore unable to address climate change in any meaningful way and are likely to exacerbate existing injustices (Swyngedouw 2007; Swyngedouw 2010; Kenis and Lievens 2014). A third critique focuses on the relationship between city greening and gentrification. This work asserts that the environmental amenities and low-carbon lifestyle that are attracting middle-class residents back to city centres are causing displacement and gentrification, raising questions as to whom the green, or low-carbon, city is for (Checker 2011; Quastel et al. 2012; McKendry and Janos 2015; Gould and Lewis 2017; Anguelovski et al. 2018; Rice et al. 2020). Though varying in focus and type of analysis, these bodies of work see city environmentalism as too little to reduce humanity's impact on the environment, as exacerbating urban inequality and gentrification, and/or failing to recognize the underlying sociopolitical causes of the climate crisis (Chap. 3). Because of these limitations in technocratic and pro-growth-focused urban responses to climate change, more radical transformation is required (Pelling 2010; Cook and Swyngedouw 2012; Kenis and Mathijs 2014). Furthermore, those who would promote such initiatives are at best naïve and at worst are striving to use public concern for the environment as a way to profit from increasing urban land values.

These arguments regarding the problems with growth-focused city environmentalism offer important insights, and expanding on them in relation to climate urbanism will undoubtedly be a part of this emerging research agenda. However, studying a place like Colorado Springs forces a rethinking of the universality of these critiques, and of their dismissal of urban greening efforts when they are undertaken explicitly to further economic growth. In a conservative political context like that of Colorado Springs, the choice is not between neoliberal urban greening and a more radical, transformative form of climate urbanism. It is between high-carbon, sprawling, car dependent business-as-usual urban development and a model that, although just as deeply committed to neoliberal urbanism, offers a modestly lower-carbon development pathway that could lay the groundwork for more significant changes. Using the city council's 2015 decision to close one of its coal-fired power plants, a move that will substantially reduce the carbon intensity of Colorado Springs' energy mix, I illustrate the need to appreciate the importance of urban climate efforts that are done without the goal of reducing carbon emissions. In this case, shutting down the power plant was pushed by local environmental activists for years with no success. It was only when the business community convinced policymakers that closing the plant would spur economic development in downtown Colorado Springs by attracting tourists and young professionals that the decision to move away from coal was made.

9.3 Business Interests and the Closing of a Coal-Fired Power Plant

As population and economic growth boom across Colorado, Colorado Springs is seeing something of a renaissance. After having no new downtown development for several decades, in 2017 around $620 million in new hotels, apartments, offices, restaurants, and public improvements had been recently completed or were in some stage of the building or planning process (Laden 2017). This included the country's only official Olympic Museum, complementing the city's new brand as "Olympic City USA" and its location as the site of the US Olympic Training Centre and the US Olympic Committee headquarters. Though much is happening, many see Martin Drake, one of two coal-fired power plants run by Colorado Springs Utilities (CSU), as the major obstacle to the full redevelopment of downtown (Paul 2019). As hundreds of coal-fired power plants have been shuttered across the United States over the past several years (Kuykendall and Cotting 2018), Martin Drake is one of the few city centre coal-fired power plants that remains in operation. Drake sits on the southwest edge of downtown, practically across the street from the Olympic Museum, tainting what is otherwise a striking view of the Rocky Mountains. That a city endowed with the natural beauty of the mountains should obscure it with a power plant, and gobble up potentially valuable downtown real estate while doing so, is, according to people working for the city and in the downtown business community, "embarrassing" and "ridiculous" (personal communications, February 2018). Reflecting this, it is local business owners and developers that have become the most effective proponents of shuttering Martin Drake.

Talk of closing Drake began in 2005, when the plant was found to be contributing to the haze surrounding Rocky Mountain National Park in northern Colorado (de Yoanna 2005). Rather than phase out the plant, CSU decided to invest over $200 million into an experimental form of Sulphur dioxide scrubbers invented by a local businessman. After the decision, serious calls to close the plant did not emerge again until 2012. This time, efforts were led by a strained, informal coalition between the local Sierra Club chapter and the city's recently elected first strong mayor (Hazlehurst 2012). The mayor himself was a developer, and one of his priorities was revitalizing Colorado Springs' downtown in order to retain the young professionals who were fleeing the city (Paul 2013). Yet, the utility board (which is also the Colorado Springs city council) resisted calls to close the plant. For example, rejecting activists' efforts to get the board to consider health impacts, environmental costs, and climate change in their calculations of the future of Drake, one utility board member stated that the plant was "paid-for, [has a] low cost to produce energy, and the money we put into it for [scrubbers] enhances it for years into the future. I don't see it as a plant that needs to go away anytime soon" (quoted in Routon 2013, unpaginated). Similarly, a number of other board members roundly rejected any arguments about Drake's contribution to climate change or about the climate and environmental benefits of moving away from coal (Hazlehurst 2015).

Around this time, however, something began to change in the city. The skyrocketing housing prices in other, more popular Colorado cities began to push development into Colorado Springs, and the business community became increasingly vocal that the time was right to close Martin Drake to take advantage of the momentum to revitalize downtown (which, just blocks from the main business strip, is scattered with vacant lots and underutilized warehouses). Letters signed by dozens of downtown business leaders were submitted to the utility board and published in the local independent newspaper (Zubeck 2014). Though environmental activists continued to show up to board meetings, as they had for years, they were now regularly joined by developers and the president of the Downtown Business Association speaking out in favour of closure. Then, in a move that surprised many, in November 2015 the board voted six to three to decommission Drake no later than 2035 (CSU Board Minutes, November 18, 2015). The downtown business community and developers were important participants in the debate leading up to the vote. As one local reporter summarized the meeting:

> [Two major local developers] both spoke out for a more timely closure of Drake, making respective cases for business and economic development as well as aesthetic concerns. [One] called Drake a 'god awful' eyesore for a community that prides itself on so much surrounding natural beauty. [The] Downtown Partnership president … also spoke for business interests, saying that routinely when she's meeting with outside interests looking to invest in our community, they ask about Drake and its future. Not being able to give an answer up to this point, she feared, showed a lack of our vision. (Stanton Anleu 2015)

The councillor who maneuvered the vote asserted that growing community pressure had led to his decision.

> "It was the community from all fronts... . We'd been getting stopped in the parking lot of the church or the post office by [developers]. We'd been getting a lot of questions about what we were going to do to help the renaissance of downtown. A number of things have been building over the last six months, and the drumbeat has been getting louder and louder." (quoted in Stanton Anleu 2015)

Another board member, who was a key unexpected vote in the passage of the measure, said, "I wasn't going to vote that way when I walked in there. ... The thing that changed my thinking was that ... if we're going to close it inevitably, what's the point of fixing [it]?" (ibid). Even as this councillor voted for the closure, however, he asserted that he did not have a problem with coal, and outright rejected "some residents' health and environmental concerns" about the plant (ibid). Other councillors focused on the importance of a fixed closure date providing certainty for long-term development planning in the community. When the vote passed, suddenly, like hundreds of coal-fired power plants across the country, Martin Drake had a closure date, even if that date was twenty years away. Since the 2015 vote, the business community has continued to pressure the city to close the plant sooner. Many on the utility board now see a 2025 closure date as a real, if ambitious, possibility, and hundreds of millions of dollars are being spent on constructing new solar arrays and integrating with the regional grid, both necessary steps to closing the power plant (Swanson 2017). When it closes, Colorado Springs' coal use and its accompanying carbon emissions and local air pollution from coal will be cut in half.

9.4 The Transformative Potential of Low-Carbon as a Growth Strategy?

The story of Martin Drake illustrates that urban policy in Colorado Springs has not even reached the most neoliberal modes of climate urbanism wherein the economic benefits of new climate-friendly industries and developments are paramount (Long and Rice 2019). This is also the case with the other climate-friendly changes that are occurring in the city such as an expansion of green transportation infrastructure and a move toward higher density brownfield redevelopment. Rather than an embrace of climate urbanism, Colorado Springs is just finally adopting the amenity-focused green urban entrepreneurialism that many cities saw as a key strategy to their revitalization decades ago (McKendry 2018). This lack of interest in explicitly environmental and climate-focused policy has been frustrating to the environmental activists who have been working for years to make Colorado Springs more sustainable. Yet, many activists have told me that they feel like climate arguments for desired changes are political non-starters at best and, in some instances, can hurt their efforts. In this political context, the economic growth arguments for moves away from carbon intensity are what finally led to changes, such as closing Martin Drake, that were largely inconceivable a decade ago.

Being a low-carbon city has come to be seen as "the sign of urban modernity and progress in the twenty-first century" (While 2014: 47). Yet questions were raised about who wins and who loses in this transition (Castán Broto and Bulkeley 2013; Bulkeley et al. 2013; While 2014; Anguelovski et al. 2016). These questions are vital. But what this discussion sometimes overlooks are the social and environmental injustices of a high-carbon city. In addition to exacerbating global climate injustices (McKendry 2016), on a local level these include transportation options that do not meet the needs of transit-dependent and low-income populations, air pollution from high-carbon energy production, and a failure to provide clean energy and efficiency measures to low-income homes, exacerbating fuel poverty. In Colorado Springs, the changes that are beginning to occur to attract businesses, tourists, and young professionals—infill development, cycling infrastructure and better public transportation, moving away from coal as a main energy source—are setting the stage for ameliorating some of the social injustices of the high-carbon city. This is the case even though these changes are not being explicitly done to further either environmental sustainability or social justice. And it is possible, though by no means guaranteed, that the economic success of Colorado Springs' green urban entrepreneurialism will create space for additional low-carbon development strategies. Some have asserted that young professionals move to city centres not only because of their amenities, nightlife, proximity to jobs (Clark 2004; Florida 2004) but because they are looking for low-carbon lifestyles (Rice et al. 2020). If this is perceived to be the case by Colorado Springs leaders, it is possible that the city will move from green urban entrepreneurialism to a more ambitious climate urbanism to promote its low-carbon credentials to these potential new residents. The explicit embrace of the reality of climate change, and of the importance of low-carbon development and industry, would be a sea change in Colorado Springs. It would open even more potential for reshaping the energy systems, infrastructure, and social relations that a socio-technical transition away from fossil fuels and environmental injustices in this deeply conservative city will require (Newell and Mulvaney 2013).

9.5 Conclusion

One of the questions driving this book is how researchers can engage with climate urbanism to make a difference in policy and practice. Colorado Springs offers two answers to this question. The first answer has to do with framing and taking advantage of the policy windows opened by non-climate-related political changes to push for climate goals. In Colorado Springs, this policy window was created by the growing pressure of the regional housing market and the belief by the new mayor and developers that downtown redevelopment was economically viable. This provided the opportunity to reframe Martin Drake as a barrier to development and put its closure on the policy agenda. The second answer is that researchers invested in climate urbanism must think about policy and practice beyond the city and beyond climate change. State-level energy policy, affordable housing policy, and regional growth trends all can shape climate urbanism. Engaging with these broader issues and scales (Chaps. 5 and 7) will be necessary to push less ambitious cities, such as Colorado Springs, toward greater climate action, and to incorporate questions of equity and justice into such transformations.

The case of Colorado Springs also raises questions for future research. In particular, it asks that a future research agenda includes urban spaces that do not articulate a climate agenda but are nevertheless being influenced by, and influencing, climate change. It asks whether such spaces and their entrepreneurial endeavours are indeed laying the groundwork for transformative climate urbanism or, alternatively, if they are merely recreating urban injustices of the past century (Chaps. 2, 3 and 4). Finally, this case calls for caution in assuming that the city and the urban are useful sites of analysis for understanding and furthering climate justice. Processes of urbanization remain deeply embedded in the global political economy. Even cities with commitments to climate action struggle to overcome the growth imperative and to make the wider transformations necessary for a just climate urbanism. In those with no such commitments, such as Colorado Springs, climate justice is unlikely to be more than an unintended side effect, and therefore unlikely to be fully realized. A vital part of the climate urbanism agenda, therefore, will be to understand the limits of the political scale of the city, and the processes of urbanization, in moving toward greater transformation.

References

Anguelovski, I., Shi, L., Chu, E., Gallagher, D., Goh, K., Lamb, Z., et al. (2016). Equity impacts of urban land use planning for climate adaptation: Critical perspectives from the global north and south. *Journal of Planning Education and Research, 36*(3), 333–348.

Anguelovski, I., et al. (2018). Assessing green gentrification in historically disenfranchised neighborhoods: a longitudinal and spatial analysis of Barcelona. *Urban Geography, 39*(3), 458–491.

Bernstein, S. (2001). *The compromise of liberal environmentalism*. New York, NY: Columbia University Press.

Bloomberg, M., & Pope, C. (2017). *Climate of hope: How cities, business, and citizens can save the planet*. New York, NY: St. Martin's Press.

Bulkeley, H. (2013). *Cities and climate change*. Abingdon: Routledge.

Bulkeley, H., Carmin, J., Castán Broto, V., Edwards, G. A. S., & Fuller, S. (2013). Climate justice and global cities: Mapping the emerging discourses. *Global Environmental Change, 23*, 914–925.

Checker, M. (2011). Wiped out by the 'greenwave': environmental gentrification and the paradoxical politics of urban sustainability. *City & Society, 23*(2), 210–229.

Clark, T. N. (2004). Urban amenities: Lakes, opera, and juice bars: Do they drive development? In T. N. Clark (Ed.), *The city as an entertainment machine* (pp. 103–140). London, UK: JAI.

Cook, I. R., & Swyngedouw, E. (2012). Cities, social cohesion and the environment: Towards a future research agenda. *Urban Studies, 49*(9), 1959–1979.

Daniels, T. (2001). Smart growth: A new American approach to regional planning. *Planning practice and research, 16*(3-4), 271–279.

de Yoanna, M. (2005). Hazy days: Colorado Springs utilities nudged over polluting plants.

Fitzgerald, J. (2010). *Emerald cities: Urban sustainability and economic development*. New York, NY: Oxford University Press.

Florida, R. L. (2004). *The rise of the creative class: And how it's transforming work, leisure, community and everyday life*. New York, NY: Basic Books.

Gibbs, D., & Krueger, R. (2007). Containing the contradictions of rapid development? New economy spaces and sustainable urban development. In R. Krueger & D. Gibbs (Eds.), *The sustainable development paradox: Urban political economy in the United States and Europe*. New York: Guilford Press.

Gould, K. A., & Lewis, T. L. (2017). *Green gentrification: Urban sustainability and the struggle for environmental justice*. New York, NY: Routledge.

Hammer, S., et al. (2011). "Cities and green growth: A conceptual framework", OECD Regional Development Working Papers 2011/08, OECD Publishing.

Hazlehurst, J. (2012, June 27). Drake: The new battleground. Independent.

Hazlehurst, J. (2015, March 25). Climate Change: Deal with It. Independent.

Janos, N., & McKendry, C. (2014). Globalization, governance, and re-naturing the industrial city: Chicago, IL and Seattle, WA. In S. Curtis (Ed.), *The power of cities in international relations*. New York: Routledge.

Jonas, A. E. G., Gibbs, D., & While, A. (2011). The new urban politics as a politics of carbon control. *Urban Studies, 48*(12), 2537–2554.

Jonas, A. E. G., & While, A. (2007). Greening the entrepreneurial city? Looking for spaces of sustainability politics in the competitive city. In R. Krueger & D. Gibbs (Eds.), *The sustainable development paradox: Urban political economy in the United States and Europe* (pp. 123–159). New York, NY: Guilford Press.

Kenis, A., & Lievens, M. (2014). Searching for 'the political' in environmental politics. *Environmental Politics, 23*(4), 531–548.

Kenis, A., & Mathijs, E. (2014). Climate change and post-politics: Repoliticizing the present by imagining the future? *Geoforum, 52*, 148–156.
Kuykendall, T., & Cotting, A. (2018). Coal plant closings double in Trump's 2nd year despite 'end of war on coal' [online]. Retrieved June 17, 2019, from https://www.spglobal.com/marketintelligence/en/news-insights/latest-newsheadlines/48671375.
Laden, R. (2017). Building boom takes shape in downtown Colorado Springs. Gazette, October 21.
Long, J., & Rice, J. L. (2019). From sustainable urbanism to climate urbanism. *Urban Studies, 56*(5), 992–1008.
McKendry, C. (2016). Cities and the challenge of multiscalar climate justice: Climate governance and social equity in Chicago, Birmingham, and Vancouver. *Local Environment, 21*(11), 1354–1371.
McKendry, C. (2018). *Greening post-industrial cities: Growth, equity, and environmental governance*. New York: Routledge.
McKendry, C., & Janos, N. (2015). Greening the industrial city: Equity, environment, and economic growth in Seattle and Chicago. *International Environmental Agreements: Politics, Law, and Economics, 15*(1), 45–60.
Newell, P., & Mulvaney, D. (2013). The political economy of the 'Just transition'. *The Geographical Journal, 179*(2), 132–140.
Paul, J. (2013). Young crowd in Colorado Springs, Part 1: What keeps young professionals away? The Gazette, September 3, 2013.
Paul, J. (2019). Colorado Springs has big plans for its downtown. But first the city must deal with the Martin Drake Power Plant. The Colorado Sun, April 4.
Pelling, M. (2010). *Adaptation to climate change: From resilience to transformation*. London: Routledge.
Quastel, N., Moos, M., & Lynch, N. (2012). Sustainability-as-density and the return of the social: The case of Vancouver, British Columbia. *Urban Geography, 33*(7), 1055–1084.
Rice, J. L., Cohen, D. A., Long, J., & Jurjevich, J. R. (2020). Contradictions of the climate-friendly city: New perspectives on eco-gentrification and housing justice. *International Journal of Urban and Regional Research, 44*(1), 145–165.
Routon, R. (2013, January 23). Hente: Committed Councilor 'til the end Independent.
Stanton Anleu, B. (2015). How and why the Colorado Springs utilities board voted for Drake closure [online]. Colorado Springs, CO. Retrieved from http://coloradosprings.com/howand-why-the-colorado-springs-utilities-board-voted-for-drake-closure/article/1563886.

Swanson, C. (2017). Downtown Colorado Springs power plant could be shuttered early, but at a price. Gazette, November 18.

Swyngedouw, E. (2007). Impossible 'sustainability' and the postpolitical condition. In R. Krueger & D. Gibbs (Eds.), *The sustainable development paradox: Urban political economy in the United States and Europe* (pp. 13–40). New York, NY: Guilford Press.

Swyngedouw, E. (2010). Apocalypse forever? Post-political populism and the spectre of climate change. *Theory, Culture and Society, 27*(2–3), 213–232.

While, A. (2014). Carbon regulation and low-carbon urban restructuring. In M. Hodson & S. Marvin (Eds.), *After sustainable cities?* (pp. 41–58). New York, NY: Routledge.

While, A., Jonas, A. E., & Gibbs, D. (2004). The environment and the entrepreneurial city: Searching for the urban 'sustainability fix' in Manchester and Leeds. *International Journal of Urban and Regional Research, 28*(3), 549–569.

Zubeck, P. (2014, July 17). 'Close Drake' proponents submit arguments. Colorado Springs Independent.

Zubeck, P. (2017, June 7). Suthers criticizes mayors opposed to Trump's decision on climate accord. Colorado Springs Independent.

Part III

The Knowledge Politics of Climate Urbanism

10

An Adaptation Agenda for the New Climate Urbanism: Global Insights

Marta Olazabal

10.1 Introduction

Urban planning is recognised as a means for climate adaptation (Carter et al. 2015): thousands of cities around the world have initiated urban adaptation planning processes. While there is an assumption that adaptation should take place through city-wide strategies (Meerow and Woodruff 2020), there is yet concern about how these processes are facilitated and which goals they pursue (Olazabal, Galarraga, et al. 2019). An unsettling concern is whether these formal science-based adaptation planning models, in the end, will be successful across contexts, as arenas for cross-scale adaptation processes are still lacking. Adaptation (or lack of it) is making visible local institutional immaturity and lack of capacities for adaptation in many cities across the world (Patterson and Huitema 2019). This suggests that, considering the diversity of approaches towards local environmental governance globally (Anguelovski et al. 2014), a key

M. Olazabal (✉)
Basque Centre for Climate Change (BC3), Leioa, Spain
e-mail: marta.olazabal@bc3research.org

V. Castán Broto et al. (eds.), *Climate Urbanism*,
https://doi.org/10.1007/978-3-030-53386-1_10

question is how adaptation is shaping cities worldwide, especially as adaptation might reinforce inequalities between the cities that are able to cope with the effects of climate change, and those that are not (Chap. 3).

Evaluating whether adaptation is progressing effectively and whether adaptation actions are or will be successful is not an easy task (Ford et al. 2015). Questions and ambiguities start with the definition of adaptation itself and with the identification of actions that can be considered as adaptation (ibid.). While this may seem a trivial question, how we define adaptation is important to understand whether local capacities and governance might support adaptation efforts and whether, how and when goals will be achieved. As vulnerabilities have a cross-sectoral nature (e.g. poverty affects mobility, health services access, community and self-care activities, education and work opportunities, housing access and thus, adaptation in multiple ways), conferring rigidity to what adaptation is will create unevenness in the long term, specially, in terms of financing (Keenan et al. 2019). In contrast, global comparisons of adaptation require specific focus and narrow terms to identify who is adapting to what, when, where and why, and to enable cross-city comparison. It is challenging to define adaptation at the city level and agreeing on adequate methods and metrics to measure progress is equally complex (Ford and Berrang-Ford 2015; Magnan 2016; Tompkins et al. 2018; Chap. 5). Even if the evaluation of (potential) outcomes is important to understand where adaptation might be leading us, the difficulties in measuring them (as most of the impacts have not happened yet) invite to focus on the evaluation of adaptation processes and outputs (Berrang-Ford et al. 2019; Hallegatte and Engle 2019). This is why most of evaluation frameworks for adaptation focus on processes and context, for example, readiness (Ford and King 2015), preparedness (Heidrich et al. 2013), barriers and enablers (Moser and Ekstrom 2010), policy credibility (Olazabal, Galarraga, et al. 2019) or the characterisation of outputs (Araos et al. 2016; Carmin et al. 2012; Le 2019; Reckien et al. 2018; Shi et al. 2015) and the evaluation of their quality (Woodruff and Stults 2016). Although there exist useful conceptual frameworks to understand what successful adaptation means (Adger et al. 2005), there has been little progress on their translation into the practical evaluation of existing initiatives in general, and, particularly, in cities. This creates further difficulties to identify what works and does not work when it comes to adaptation. Likewise, it makes it difficult to develop transferable approaches to: (1) monitor specific initiatives or programmes; (2) measure success of implemented interventions; (3) identify hot and cold spots of adaptation activity; and, (4) provide credible adaptation models (Olazabal, de Gopegui, et al. 2019).

This chapter offers a critical reflection on the results of a global urban adaptation tracking study covering 136 coastal cities around the world (Olazabal, de Gopegui, et al. 2019). The study was originally developed with the aim of documenting adaptation-related initiatives in coastal cities worldwide and of building a reference baseline of what is happening on the ground. To reflect the role of multilevel governance in local responses to climate change, the study captures adaptation initiatives across different tiers of governments. The study makes transparent use of an adaptation-documenting protocol (following Tompkins et al. 2018) that can be replicated elsewhere. This chapter summarises the most important results of this study, presented in more length in Olazabal, de Gopegui, et al. (2019), and discusses them in relation to climate urbanism. The chapter offers reflections on how adaptation—as city-wide strategies—feed into new forms of urbanism in response to climate change, and how the emergence of climate urbanism might force us to re-think adaptation in diverse ways.

10.2 Data and Methods

Context and Framing

Olazabal, de Gopegui, et al. (2019) documented government-led (top-down) adaptation initiatives in coastal city-regions worldwide. In an effort to go beyond city administrations and look at multilevel governance to understand climate urbanism, the paper examined different tiers of government whose strategies had an impact on coastal cities. From November 2018 until April 2019, we collected and analysed 226 national, regional and local adaptation-related policies affecting the 136 largest coastal port cities worldwide—a set of cities for which coastal-related risks have been widely studied (Abadie et al. 2017; Hallegatte et al. 2013; Hanson et al. 2011). These coastal cities concentrate around 700 million inhabitants (approximately 10% of the world population) and the 68 countries where they are located, contain over 6 billion people (almost 82% of the world population). To develop the document protocol, we followed the four-step stock-taking approach designed by Tompkins et al. (2018): (a) obtaining consensus on the objectives of adaptation; (b) agreeing on the sources of evidence; (c) agreeing on the search method; and (d) categorising different approaches to adaptation. The documenting protocol consisted of a search protocol, a selection protocol and a characterisation protocol (see Fig. 10.1) with the final aim of enhancing

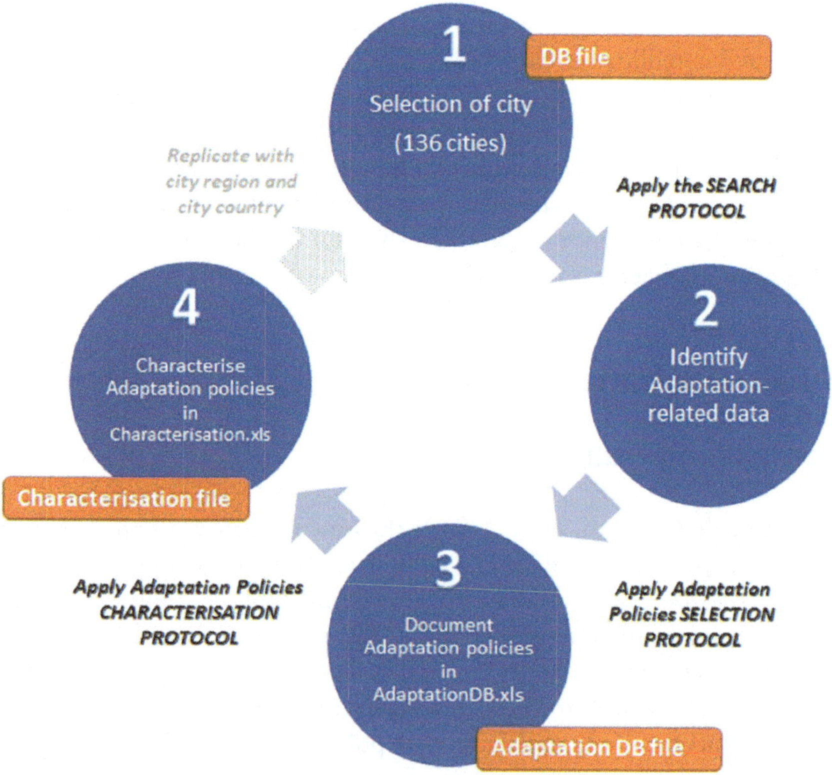

Fig. 10.1 Documenting protocol (Olazabal, de Gopegui, et al. 2019)

replicability and comparability of the results, thereby increasing the transparency of the process (ibid.).

Adaptation Objectives and Sources of Evidence

The 226 documents collected reflect explicit public sector action on climate change adaptation (Dupuis and Biesbroek 2013). These policies are both adaptation-focused policies, as well as those integrating adaptation-related objectives and actions demonstrating intention to reduce risks from climate change, including those intended to reduce, specifically, coastal risks. Collected policies date from 2006 until early 2019. These policies are the latest information available on governmental adaptation plans. We did not consider public policies that incentivise adaptation but are not motivated by climate change (e.g. an ecological restoration program). The data collected includes documents such as strategies, programmes and plans (not laws or regulations). The study aims to provide a state-of-the-art assessment of government adaptation goals and actions and, more specifically, to collect evidence on the alignment of current action to future risks. For this reason, the review assumed that adaptation policy should be evidence-based, using risk and vulnerability assessments, future climate change scenarios and adaptation options assessments.

Search Methods and Categorisation

We collected documents through public websites searches and/or provided by city officials. We used an Internet search engine (Google) and verified information with onsite experts and public officials (more details on the methods can be found in Olazabal, de Gopegui, et al. 2019). Selected documents use explicit terms related to adaptation in their title (i.e. climate change adaptation) or have a preamble or specific sections related to adaptation where the motivation is clearly stated. Furthermore, they contain a definition of measures of (or actions towards) adaptation. Selected policies focus on climate change adaptation, climate change mitigation and adaptation, coastal-related management and flood risk management policies that consider adaptation. Following data collection, policies were characterised according to: (1) basic policy data (name, responsible authority, policy scale, publishing year, approval year, policy lifespan), (2) type of policy and (3) adaptation contents including climate information. We identified 41 adaptation policy characteristics in total. The data collected and metadata (including search method for each of these characteristics) can be found in the supplementary material of Olazabal, de Gopegui, et al. (2019).

We classified policies according to five geographical scales:[1] national, state, regional, metropolitan and city scales. The national level consists of country-specific policies. State policies are those developed by smaller divisions of federal countries (e.g. Brazil, India, USA, Germany). Regional policies correspond to provinces and land areas under the responsibility of a regional administrative body. A metropolitan policy may cover various municipalities including the city proper. City-level policies refer to the administrative boundaries (city proper) of the local administration. Example: Country: Spain, Region: Catalonia, Metropolitan area: Área Metropolitana de Barcelona, City: Barcelona. Through the application of the documenting protocol (Fig. 10.1), seven main types of adaptation-related policies have been identified: climate adaptation policies (A), climate change policies including both mitigation and adaptation objectives (A/M), coastal management policies (C), coastal adaptation policies (A;C), disaster risk reduction policies (DRR) and disaster and adaptation policies (A;DRR). The rest are sustainability, resilience and master plans which are grouped as 'Others'. C and A;C differ in that the latter policies are normally specifically developed to articulate action on coastal adaptation, while the former (C) are coastal management plans that contain measures addressed to increase adaptation. Similarly, A;DRR policies jointly articulate adaptation and disaster management, while DRR policies have less planning for adaptation.

[1] In some cases, these 5 scales are further grouped in three scales for simplification: national, regional (state + regional) and local (city + metropolitan).

10.3 Results and Implications for Climate Urbanism

Where Is Adaptation Policy Activity Located Globally?

We found that, overall, adaptation policy is reasonably distributed across national, regional/state and city/metropolitan scales. However, there are important differences in the adaptation efforts undertaken by different tiers of government across world regions. Adaptation planning in Asia, Africa and Latin America, for example, is mostly dominated by national and local (city/metropolitan) plans. This means that, in these countries, national governmental policies have the potential to influence climate adaptation in cities. There are exceptions in these world regions. For example, in Japan and South Africa, regional authorities are active actors in adaptation planning processes. The world region where national governments are more active is Africa (see Table 10.1), while, for example, state governments are most active in the USA. Our city sample is dominated by Asian cities (~40%) as they concentrate the largest coastal port

Table 10.1 Number of countries and cities per world region with no planning (Olazabal, de Gopegui, et al. 2019)

World regions	Total countries analysed	Total cities analysed	Total number adaptation policies (% of total adaptation policies found)	Number of countries with no planning[a] (% of total countries in world region)	Number of cities with no planning[b] (% of total cities in world region)
Africa	16	19	26 (11.5%)	1 (6.2%)	16 (84.2%)
Asia	22	54	55 (24.3%)	5 (22.7%)	39 (72.2%)
Europe	14	17	50 (22.1%)	2 (14.3%)	5 (29.4%)
Latin America	12	21	32 (14.1%))	2 (16.7%)	13 (61.9%)
North America	2	19	43 (19.0%)	1 (50.0%)	5 (26.3%)
Oceania	2	6	20 (8.8%)	0 (0.0%)	1 (16.7%)
Total	68	136	226 (100%)	11 (8.1%)	79 (58.1%)

[a]At national level
[b]At local level (city and/or metropolitan policies)

cities worldwide. Consequently, our sample of policies is dominated by Asian policies (24%) followed by Europe and North America. Nevertheless, the highest rate of adaptation activity relative to the number of cities analysed is found in Europe and North America, coinciding with the highest activity of regional governments.

At the local scale, there is a significant gap in adaptation policy in certain regions of the world. In Africa, the vast majority of coastal cities in the sample (~85%) do not have local government-led adaptation policy, followed by Asia and Latin America (see Table 10.1 and Fig. 10.2). The percentage of cities without adaptation plans increases in these continents when the most developed countries are removed from consideration, that is, South Africa, Brazil, South Korea and Japan. For example, only 3 of the 19 African cities analysed have formal local planning (Dakar, Durban and Cape Town). While Dakar has a brand-new urban resilience strategy with little weight on adaptation planning, Durban and Cape Town have larger experience in climate change policy and adaptation planning, and, therefore, practical knowledge and a well-established adaptation community. Our findings, though, suggest that national policy does not trickle down to regions or cities in the developing world,

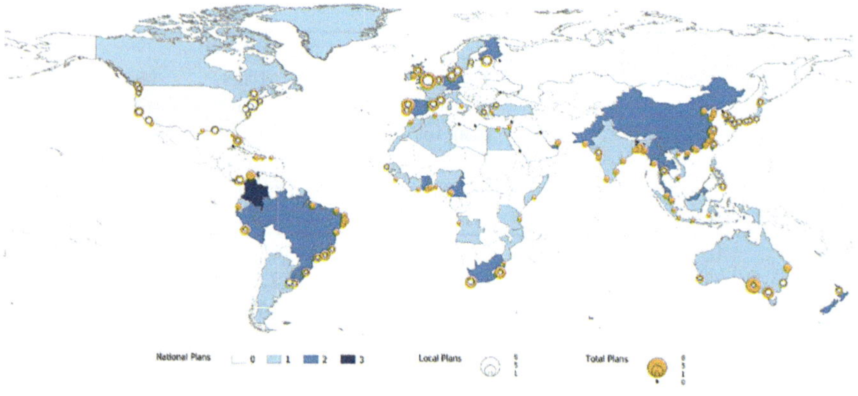

Fig. 10.2 Number of adaptation policies at national, regional and local levels per city (Olazabal, de Gopegui, et al. 2019). Orange bubbles indicate the total number of plans; white bubbles indicate the total number of local (city and metropolitan) plans. Sampled countries are shaded gradually indicating the number of national policies

highlighting an absence of local government-led action. Reasons for this include an absence of legal frameworks and administrative, financial, political capacities to plan for adaptation in a top-down formal way. At the global level, 58% of coastal cities over 1 million inhabitants do not have any local adaptation planning in place. Moreover, eleven of these cities—Kuwait City (Kuwait), Banghazi (Libya), Beirut (Lebanon), Jiddah (Saudi Arabia), Havana (Cuba), Maracaibo (Venezuela), N-ampo (Dem. People's Republic of Korea), Odessa (Ukraine), St. Petersburg (Russia), Tripoli (Libya)—are located in regions and countries with no adaptation activity. Most of these cities are located in developing regions where, as suggested by this analysis, adaptation has not progressed in any form across scales. The USA is the country without national adaptation framework with the most populated coastal cities. In general, most adaptation policies are delivered in richer areas both in developed or developing world regions.

What Is the Morphology of Adaptation?

Although, sustainability, resilience and other planning tools such as master plans (namely, 'others') are used at the local level to articulate adaptation (~30%), two-thirds of documents are either adaptation policies (A) or policies that integrate adaptation and mitigation (A/M). This might wrongly lead to suggest that adaptation is still siloed in strategic climate planning and that, globally, it has not yet mainstreamed in other kinds of sectoral policies. However, the protocol we developed does not capture, for example, sectoral policies such as those on transport, biodiversity or social protection which could be already mainstreaming adaptation criteria into their planning. Our database thus cannot be used to extract conclusions around mainstreaming and adaptation institutionalisation (Chap. 5). This stresses the importance of providing transparent information about how adaptation is studied to inform future climate urbanism research, as mapping out the mainstreaming adaptation efforts will also require to look beyond climate policies. Most adaptation policies collected in our sample were published after 2014 (~70%). As the protocol was developed to collect the last known adaptation policies in force, this suggests that adaptation activities could be undergoing a process of implementation. Yet, almost half of the policies published after 2014 are first generation plans, suggesting that adaptation might be in a state of immaturity across contexts and scales. Collected policies are published documents which are only approved in 40% of the cases. However, the importance of policy approval processes differs across countries. For example, Spanish planning requires approval of governmental bodies to recognise the commitment and responsibility of the authority for the implementation of the plan. In the USA, in contrast, while the approval rates are good, such bureaucratic processes are not common suggesting approval does not link to governmental responsibility and thus, cannot be used as a global indicator for implementation progress or effectiveness. Policies have a strategic (~40%) or tangible (~60%) nature in the cases where more concrete measures are included. Local and regional adaptation-related policies tend to be more concrete, whilst half of national policies are more strategic. We have also used Lesnikowski et al. (2011, 2013) framework (R: recognition, G: groundwork and A: action) to provide an overall perception of the type of measures contained in each document. Given that our protocol has not captured policies that were limited to the recognition of the need to adapt (R), all captured policies included measures to increase knowledge on impacts, risks, and benefits (G) and measures directed to change key areas such as infrastructure, legislation, and institutions (A).

What Is the Focus of Adaptation Policies Worldwide?

Although the database published in Olazabal, de Gopegui, et al. (2019) does not go deeper into which kinds of climatic impacts are being addressed in each policy, it provides a general overview of their sectorial focus. The analysis gives a flavour of the differences and similarities in adaptation activity across policy scales and world regions. Globally, the most frequent focus of adaptation measures is on governance, coast and water, disaster risk reduction, ecosystems and urbanisation (70–80%). National policies deal more often with governance, health, ecosystems, energy, agriculture and food issues. Metropolitan plans focus more on waste (25% of metropolitan plans) than city plans (7% of city plans).

Overall there is a good cover of adaptation directed to ecosystems across scales, but, remarkably, city plans rarely focus on ecosystems (~58%). Urban green solutions are found in cities that are strongly embracing new climate urbanism agendas, as these are seen as flexible low regret options for adaptation planning where multiple interests meet (Frantzeskaki et al. 2019), often dismissing equity and justice considerations (Anguelovski et al. 2019). Our data suggests that adaptation in cities does not focus on increasing the resilience of urban ecosystems, on evaluating their health and on maximising their role in urbanism but rather, adaptation is centred on the use cities can make of urban ecosystems as urban adaptation furniture. Remarkably, 66% of national policies deal with adaptation in urban planning, urban infrastructures or the built environment, while only ~30% of local policies focus on them. Regional (24%) and state (12%) policies have less coverage of the urban built environment. This evidences the need to go beyond local administrative boundaries and consider multilevel governance when contextualising the emergence of climate urbanism (see Chaps. 5 and 7).

We found significant regional variations in the focus of policies, possibly reflecting different adaptation needs or progress in each world region. Almost all policies in Asia cover governance issues, in contrast to around half of adaptation policies in North America or Oceania. Africa addresses the finance and waste sectors more than other regions. Notably, North American policies have the least focus on ecosystems and health (around 30%) in contrast to Asia (between 90 and 100%). Transport is most often covered in Oceania while disaster risk management is prevalent in Asia (both ~95%). We cannot speculate why this is, as it might be influenced by the typology of cities selected in the sample (large coastal cities) or by adaptation and planning cultures in each of these world regions (and within them, in each policy scale), but this evidence base allows us to start to ask appropriate questions about the sectoral progress of adaptation.

Are Adaptation Policies Followed-Up On?

The analysis suggests that not all adaptation policies are implemented. According to government reports (particularly, according to the dates associated to policy lifetimes), 80% of the policies should be in the process of implementation and ~8% should be already implemented (the rest is too recent, >2018). Actual evidence of implementation was found in only half of these cases (implementation information was collected in official websites and progress reports searching for information showing that at least one measure was being implemented). There is evidence that some old policies never progressed to an implementation stage. More worryingly, over two-thirds of identified policies fail to include a budget for full or partial implementation of the defined measures. This evidence a lack of resources for implementation and raises concerns around the adequacy of the search method across developed versus developing contexts (a large-n study such as this one has limitations on data search; we assumed that implementation progress should be publicly reported). Internet and official websites are marketing tools and data storage sites which are not widespread in developing regions. Moreover, resources, capacity and tools are required across different geographical contexts to enable the emergence of adaptation regimes that address climate urbanism. It is important to understand what successful adaptation means across diverse contexts, particularly with regards to the emphasis on outputs rather than on processes for instance. In this regard, adaptation monitoring and evaluation frameworks are important to institutionalise learning processes and to understand factors that may affect adaptation policy implementation and delivery. Promisingly, MERs (monitoring, evaluation and reporting systems) are proposed in almost 90% of the policies. This, however, requires further investigation as preliminary research shows that MER systems and learning mechanisms are not adequately defined in current policies.

An Evidence Base for Climate Urbanism

The scholarly literature argues that climate planning requires a 'strong fact base' and that it should incorporate "empirical data on current conditions (GHG inventory or vulnerability assessment), future projections, and modelled impacts to ensure strategies are well informed" (Meerow and Woodruff 2020: 41). Our analysis shows that, often, empirical data and climate projections are not incorporated into climate planning. Moreover, climatic information, projections and assessments are not used to make decisions on adaptation. The question emerging is twofold: (1) is science producing climate information usable for adaptation decision-making, and (2) is the current model of evidence-based generation for adaptation applicable across different geographical contexts and policy scales?

In more than half of the cases, there is a quantitative or qualitative vulnerability analysis of population and/or built and natural assets. Most national documents (~70%) assess vulnerabilities to climate change compared to only ~40% of regional plans. The vast majority of policies labelled as 'adaptation' (70–100%) include vulnerability assessments, in contrast to those types of policies that are not exclusively developed to articulate adaptation such as Coastal (C), and sustainability, resilience and master plans ('Others') (~30%). Risk assessment are lacking in almost half of the policies in this database, most frequently in sustainability/resilience policies and in regional policies. European policies include risk assessments most often (70%), while those are rare in African plans (~35%). Whilst most policies that include a risk assessment have also used climate scenarios, a significant percentage (~25%) have not, suggesting they might be based only on current or past vulnerabilities. Our analysis shows that future socio-economic projections are rarely incorporated in adaptation policies. The inclusion of this type of data is mostly limited to population projections. Far more policies consider climate projections (~76%). Sustainability, resilience and master plans ('Others') use them the least (~47%) which reflects the broad scope and strategic nature of these plans with less scientific weight. In general, ~85% of the policies do not justifiably document decisions taken on adaptation with vulnerability/risk assessments or climate projections. Looking particularly at those that do generate risk assessments, only 24% of the policies aligned their proposed actions to identified risks. However, state and city policies (and coastal, C, coastal adaptation, A;C and disaster risk reduction, DRR policies) are most likely to propose actions aligned to climatic risks or scenarios. Interestingly, this model of evidence base for adaptation planning is more common in North America, which, according to our analysis, is the world region where policies most frequently propose adaptation actions stating the specific risks and/or scenarios that they are intended to address.

10.4 Conclusion

This chapter has discussed the results of a global cross-scale assessment of public sector action on climate adaptation. Beyond discussing the implications of adaptation progress worldwide, the chapter helps evaluating what this means for climate urbanism. Adaptation planning is for the most part lacking and implementation rarely accomplishes original expectations. Several reflections can be extracted from this data. On the one hand, formal adaptation planning is relatively recent and has no clear pathway to implementation. This indicates that experience in understanding the effects of planned actions is yet to emerge, and thus, that the success or effectiveness of current adaptation policies cannot be measured. There are important geographical differences in adaptation policy activity, in the forms of adaptation chosen, in the sectorial focus of adaptation efforts and in how climate information is incorporated into adaptation strategies. These divergences in adaptation across polices and scales of governance provide sufficient evidence of the need to adopt a multilevel governance approach to understand climate urbanism. On the other hand, the resulting landscape of adaptation policy activity shows that evidence-based adaptation planning (Meerow and Woodruff 2020) is concentrated in high-income regions with capacity for action. This points to a mismatch between the understanding of what adaptation planning should be (ibid.) and the forms of adaptation planning that are actually delivered in practice. This incongruence leads us to interrogate the usability of climatic information in current policy processes, the need for institutional innovation, and forces us to think about adaptation in alternative ways (beyond the plan itself). The analysis presented here suggests that we are in an era of technocratic adaptation in climate urbanism, where scientific framing might be a barrier for adaptation progress (see also Chap. 8). The lack of uniform policy progress in public sector adaptation across worldwide regions calls for further evaluations to generate new thinking about different approaches to urban climate adaptation, including new arenas, new data, new partnerships and new policy tools.

Acknowledgements This study is part of the project CLIC (Are Cities Prepared for Climate Change? http://clic.bc3research.org/) supported by AXA Research Fund (Grant Agreement No. 4771), by the Spanish Government (Grant Agreement No. IJCI-2016-28835 and the María de Maeztu excellence accreditation 2018-2022 Ref. MDM-2017-0714); and by the Basque Government (BERC 2018-2021 program).

References

Abadie, L. M., Galarraga, I., & de Murieta, E. S. (2017). Understanding risks in the light of uncertainty: Low-probability, high-impact coastal events in cities. *Environmental Research Letters, 12*, 014017.

Adger, W. N., Arnell, N. W., & Tompkins, E. L. (2005). Successful adaptation to climate change across scales. *Global Environmental Change Part A, 15*, 77–86. agenda. Energy Research and Social Science 44, 304–311.

Anguelovski, I., Chu, E., & Carmin, J. (2014). Variations in approaches to urban climate adaptation: Experiences and experimentation from the global South. *Global Environmental Change, 27*, 156–167.

Anguelovski, I., Connolly, J.J.T., Pearsall, H., Shokry, G., Checker, M., Maantay, J., Gould, K., Lewis, T., Maroko, A., & Roberts, J.T. (2019). Opinion: Why green "climate gentrification" threatens poor and vulnerable populations. *PNAS ,116*, 26139–26143. https://doi.org/10.1073/pnas.1920490117.

Araos, M., Berrang-Ford, L., Ford, J. D., Austin, S. E., Biesbroek, R., & Lesnikowski, A. (2016). Climate change adaptation planning in large cities: A systematic global assessment. *Environmentl Science Policy, 66*, 375–382.

Berrang-Ford, L., Biesbroek, R., Ford, J. D., Lesnikowski, A., Tanabe, A., Wang, F. M., Chen, C., Hsu, A., Hellmann, J. J., Pringle, P., Grecequet, M., Amado, J.-C., Huq, S., Lwasa, S., & Heymann, S. J. (2019). Tracking global climate change adaptation among governments. *Nature Climate Change, 9*, 440.

Carmin, J., Nadkarni, N., & Rhie, C. (2012). Progress and Challenges in Urban Climate Adaptation Planning. Massachusetts Institute of Technology, Massachusetts, US.

Carter, J. G., Cavan, G., Connelly, A., Guy, S., Handley, J., & Kazmierczak, A. (2015). Climate change and the city: Building capacity for urban adaptation. *Progress in Planning, 95*, 1–66.

Dupuis, J., & Biesbroek, R. (2013). Comparing apples and oranges: The dependent variable problem in comparing and evaluating climate change adaptation policies. *Global Environmental Change, 23*, 1476–1487.

Ford, J. D., Berrang-Ford, L., Biesbroek, R., Araos, M., Austin, S. E., & Lesnikowski, A. (2015). Adaptation tracking for a post-2015 climate agreement. *Nature Climate Change, 5*, 967–969.

Ford, J.D., & Berrang-Ford, L. (2015). The 4Cs of adaptation tracking: consistency, comparability, comprehensiveness, coherency. *Mitig Adapt Strateg Glob Change, 21*, 839–859. https://doi.org/10.1007/s11027-014-9627-7.

Ford, J. D., & King, D. (2015). A framework for examining adaptation readiness. *Mitig Adapt Strateg Glob Change, 20*, 505–526.

Frantzeskaki, N., McPhearson, T., Collier, M. J., Kendal, D., Bulkeley, H., Dumitru, A., Walsh, C., Noble, K., van Wyk, E., Ordóñez, C., Oke, C., & Pintér, L. (2019). Nature-based solutions for urban climate change adaptation: Linking science, policy, and practice communities for evidence-based decision-making. *BioScience, 69*, 455–466.

Hallegatte, S., & Engle, N. L. (2019). The search for the perfect indicator: Reflections on monitoring and evaluation of resilience for improved climate risk management. *Climate Risk Management, 23*, 1–6.

Hallegatte, S., Green, C., Nicholls, R. J., & Corfee-Morlot, J. (2013). Future flood losses in major coastal cities. *Nature Climate Change, 3*, 802–806.

Hanson, S., Nicholls, R., Ranger, N., Hallegatte, S., Corfee-Morlot, J., Herweijer, C., & Chateau, J. (2011). A global ranking of port cities with high exposure to climate extremes. *Climatic Change, 104*, 89–111.

Heidrich, O., Dawson, R., Reckien, D., & Walsh, C. (2013). Assessment of the climate preparedness of 30 urban areas in the UK. *Climatic Change, 120*, 771–784.

Keenan, J. M., Chu, E., & Peterson, J. (2019). From funding to financing: Perspectives shaping a research agenda for investment in urban climate adaptation. *International Journal of Urban Sustainable Development, 11*, 297–308.

Le, T. D. N. (2019). Climate change adaptation in coastal cities of developing countries: Characterizing types of vulnerability and adaptation options. Mitig Adapt Strateg Glob Change. https://doi.org/10.1007/s11027-019-09888-z.

Lesnikowski, A. C., Ford, J. D., Berrang-Ford, L., Barrera, M., & Heymann, J. (2013). How are we adapting to climate change? A global assessment. *Mitig Adapt Strateg Glob Change, 20*, 277–293. https://doi.org/10.1007/s11027-013-9491-x.

Lesnikowski, A. C., Ford, J. D., Berrang-Ford, L., Paterson, J. A., Barrera, M., & Heymann, S. J. (2011). Adapting to health impacts of climate change: A study of UNFCCC Annex I parties. *Environ. Res. Lett., 6*, 044009. https://doi.org/10.1088/1748-9326/6/4/044009.

Magnan, A.K. (2016). Climate change: Metrics needed to track adaptation. *Nature, 530*, 160–160. https://doi.org/10.1038/530160d.

Meerow, S., & Woodruff, S. C. (2020). Seven principles of strong climate change planning. *Journal of the American Planning Association, 86*, 39–46.

Moser, S. C., & Ekstrom, J. A. (2010). A framework to diagnose barriers to climate change adaptation. *Proceedings of the National Academy of Sciences, 107*, 22026–22031.

Olazabal, M., Galarraga, I., Ford, J., Sainz de Murieta, E., & Lesnikowski, A. (2019a). Are local climate adaptation policies credible? A conceptual and operational assessment framework. *International Journal of Urban Sustainable Development, 11*, 277–296.

Olazabal, M., de Gopegui, M. R., Tompkins, E. L., Venner, K., & Smith, R. (2019b). A cross-scale worldwide analysis of coastal adaptation planning. *Environmental Research Letters, 14*(12), 124056.

Patterson, J. J., & Huitema, D. (2019). Institutional innovation in urban governance: The case of climate change adaptation. *Journal of Environmental Planning and Management, 62*(3), 374–398.

Reckien, D., Salvia, M., Heidrich, O., Church, J. M., Pietrapertosa, F., De Gregorio-Hurtado, S., D'Alonzo, V., Foley, A., Simoes, S. G., Krkoška Lorencová, E., Orru, H., Orru, K., Wejs, A., Flacke, J., Olazabal, M., Geneletti, D., Feliu, E., Vasilie, S., Nador, C., Krook-Riekkola, A., Matosović, M., Fokaides, P. A., Ioannou, B. I., Flamos, A., Spyridaki, N.-A., Balzan, M. V., Fülöp, O., Paspaldzhiev, I., Grafakos, S., & Dawson, R. (2018). How are cities planning to respond to climate change? Assessment of local climate plans from 885 cities in the EU-28. *Journal of Cleaner Production, 191*, 207–219.

Shi, L., Chu, E., & Debats, J. (2015). Explaining progress in climate adaptation planning across 156 U.S. municipalities. *Journal of the American Planning Association, 81*(3), 191–202.
Tompkins, E. L., Vincent, K., Nicholls, R. J., & Suckall, N. (2018). Documenting the state of adaptation for the global stocktake of the Paris Agreement. *Wiley Interdisciplinary Reviews: Climate Change, 9*, e545.
Woodruff, S. C., & Stults, M. (2016). Numerous strategies but limited implementation guidance in US local adaptation plans. *Nature Climate Change, 6*, 796–802.

11

The New Climate Urbanism: A Physical, Social, and Behavioural Framework

Luna Khirfan

11.1 Introduction

Climatic uncertainty, entrenched power structures, established priorities, and financial and logistical difficulties make it challenging to adapt cities' built form to climate change (Kates et al. 2012; Pelling et al. 2015). While most of the climate adaptation discourse underscores social learning, for instance, through governance processes (e.g. Burch et al. 2014; Castán Broto et al. 2019), discussions on transforming urban physical structures remain, for the most part, underexplored (McCormick et al. 2013). Scholarly debates on climate urbanism have failed to seriously discuss how cities' built form could adapt to a changing climate through land use planning and urban design. In this chapter, I explore the potential of Kevin Lynch's (1981) Good City Form theory to inform transitions towards a truly transformative climate urbanism. Specifically, this chapter asks: how can this framework help us analyse and inform the

L. Khirfan (✉)
University of Waterloo, Waterloo, ON, Canada
e-mail: luna.khirfan@uwaterloo.ca

V. Castán Broto et al. (eds.), *Climate Urbanism*,
https://doi.org/10.1007/978-3-030-53386-1_11

physical, social, and behavioural changes needed for a transformative climate urbanism?

Lynch's theory is relevant to climate action in cities. Akin to the adoption of climate adaptation and mitigation as universal normative objectives in global and urban climate policy discourses, Lynch's Good City Form is "a general normative theory" that is based on universal human values. It includes five + two performance dimensions, namely fit, vitality, access, control, and sense along with the overarching objectives of efficiency and justice. These dimensions are compatible with any cultural context while simultaneously attaining local particularisms that parallel the complex local geographies of climate vulnerabilities. Furthermore, this theory captures Wirth's (1938) complex physical, social, and behavioural lenses when it defines urbanism as

> the spatial arrangement of persons doing things, the resulting spatial flows of persons, goods, and information, and the physical features which modify space in some way significant to those actions [...] The cyclical and secular changes in those spatial distributions, the control of space, and the perception of it [...]. (Lynch 1981: 48)

In this chapter, after briefly reviewing Lynch's five + two performance dimensions, I will explore how this framework can help us analyse and understand the different forms of climate urbanism we see emerging in different cities, namely the reactive, entrepreneurial, and transformative modalities of climate urbanism highlighted in this book's introduction and conclusion. In doing so, I hope to show how insights from urban design can contribute to future research on climate urbanism.

11.2 Understanding Climate Urbanism Through Lynch's Framework

Fit and vitality are the two dimensions of Lynch's framework that most obviously connect to climate urbanism, from an urban design perspective. Fit primarily means adaptability. It is relevant to climate adaptation efforts because it focuses on the adaptability of the urban form itself and on how urban forms facilitate the adaptability of human behaviours to a changing climate. Fit—as adaptability—is concerned with future action, changes, and uncertainty (see also McHarg 1992). Fit relates to characteristics of manipulability and reversibility that are relevant to climate mitigation, adaptation, and resilience objectives. Manipulability entails incremental changes to the urban form or to behaviours, while reversibility can be defined as the ability to "retrace […] to an earlier state […] to undo a mistake (or even to repeat it, if one wishes)" (Lynch 1981: 172). Vitality relates to human health, biology, and survival in cities. Vitality's key characteristics are that of sustenance, consonance, and safety in relation to access to resources (e.g. access to food, energy, water, air, and waste systems), harmony with human biology (e.g. the regulation of temperatures), and risk reduction (e.g. the control of climate risks and access to shelters and evacuation routes) (Lynch 1981).

Together fit and vitality entail design principles for different forms of climate urbanism because they attend to human-nature interactions, for instance, through land use planning. Fit and vitality can inform transformative climate urbanism where land use planning strategies stem from attempts at synergies between human needs and healthy ecosystems rather than the conventional streets, blocks, and buildings patterns. Staten Island is a case in point where a map overlay of Hurricane Sandy's flood surge impacts and of the land uses based on fit and vitality (proposed by Ian McHarg in 1969) reveals that, had these design principles been taken into consideration, the loss to human life and property due to storm surge would have been considerably less (Wagner et al. 2016). Furthermore, attending to cities' fitness and vitality transcends traditional ecosystem services approaches to human-nature interactions, which usually focus on safety and aesthetics and which have been shown to generate climate gentrification (Keenan et al. 2018). In contrast, the ecological wisdom of fitness and vitality in Lynch's framework seeks to address spatial inequalities and climate change in a transformative way, combining social values and spatial devices that also allow experimentations with old and new technologies (Castán Broto et al. 2019; Lynch 1981: 121; Newman 2016). However, fit and vitality can also be deployed as principles that support reactive forms of climate urbanism, with the objective to "moderate or avoid harm," and initiatives that correspond to more entrepreneurial forms of climate urbanism hoping to "exploit beneficial opportunities" (Mach et al. 2014: 118) brought about by climate change.

Two other dimensions of Lynch's framework, namely spatial control and access, are also relevant to these modalities of climate urbanism. Control entails the regulation of behaviours and actions within space. Access refers to the optimal, as opposed to the maximum, distributive access[1] to other people, services, resources (food, water, energy), natural environment (open spaces, shelters, wastelands, and symbolic places), information, and activities. Good access hinges upon the diversity of things being accessed, the equity of access given to different groups in society, and different groups' ability to control access to different urban resources. Long and Rice (2019) have argued that control and access constitute essential features of climate urbanism. The selective and reactive policies in our transition towards climate urbanism ensue in controlled access to climate resilient lifestyles, through controlling GHG emissions, upscaling green infrastructure, and creating climate-proof enclaves (Rice et al. 2020). These policies worsen the exclusion of marginalized and vulnerable groups from risk-free spaces.

Access and control are key features of contemporary "climate proof" urban design strategies, particularly when those focus on improving the accessibility of urban space and controlling carbon emissions. For example, transit-oriented developments modify transit systems to be less carbon-intensive by enhancing the routes and capacity of public transportation; by promoting public transportation, cycling and walkability; and by bringing origins and destinations closer through the promotion of compact urban forms (Calthorpe 1993; Ellin 1999; Lynch 1981). Large-scale infrastructure projects in cities of the global North have been reactive for the most part, centred around carbon control and protecting the circulation of urban populations, goods and services. The question of control over space is strongly related to social and behavioural changes through responsibility. The explicit or implicit ownership of space shapes how individuals can claim their spatial rights, namely the right of presence, use, action, appropriation, modification, and disposition (Lynch

[1] Optimization may be achieved with minimum time, effort, and energy and with maximum safety (Lynch 1981).

1981). When these rights are limited (to various degrees) through the privatization of housing and public or green spaces, for instance, or through increased surveillance, conflicts can occur. Conflicts over spatial control can act as triggers to social mobilization for climate action and other rights to the city, possibly catalysing collective action for transformative climate urbanism.

Lynch's framework also assesses the performance of urban forms in relation to "sense", which refers to the connections between the physical elements of a place (buildings, streets, and open spaces), the activities occurring within these, and the perception (e.g. meanings) and cognition (e.g. orientation) of a place. How we "sense" cities can inform reactive forms of climate urbanism centred on risk and hazard mapping to facilitate self-orientation during climate emergencies so as to identify optimal evacuation routes and access to services. Furthermore, sense brings to the fore local particularisms through identity and symbolic significance. The sense of place identity, or its genius loci (Norberg-Schulz 1980), stems from a combination of tangible and intangible elements, including local values, events, and feelings that manifest through local activism and engagement with the public realm (Jivén and Larkham 2003). Symbolic significance refers to the holistic meaning of urban forms and the ability of urban dwellers to be cognizant of their environment in relation to their central beliefs (Lynch 1981). These are important considerations when thinking about how human-nature relations are reconfigured through different modalities of climate urbanism (see Chap. 2).

11.3 Situating Climate Urbanism(s): Lessons from Charlottetown, Amman, Negril, and Zürich

Through examples from Charlottetown, Amman, Negril, and Zürich, I discuss how the different dimensions of Lynch's framework can help understand the emergence and impact of climate urbanism in very different cities. In Charlottetown, the capital of Prince Edward Island, a province in the Canadian Maritimes, strong connections exist among the city's fitness and vitality. Data from charrettes (interactive, drawing-oriented community participation activities) and from historic archival research reveal that the extensive acquisition of land from water increased the city's vulnerability to climate impacts. First, land was reclaimed from a watershed consisting of a stream and Governor's Pond—currently, the former runs in underground culverts while the latter is a parking lot. In addition, one of Charlottetown's former mayors eliminated the city's network of bioswales. Bioswales are naturalized channels that help with rainwater management and the recharging of underground water aquifers. These measures led to the proliferation of impervious surfaces throughout the city and to severe inland inundation from stormwater runoff (further exacerbated during extreme weather events). They also precluded the replenishment Charlottetown's aquifer which, combined with salt intrusion and cruise ship water extraction, render the city's water sustenance vulnerable. Transformative design interventions focused on reversibility would overcome these risks by reintroducing the bioswales, Governor's Pond, and stream (Fig. 11.1)[2] (Khirfan and El-Shayeb 2019;

Fig. 11.1 A composite showing the potential of fit's manipulability and reversibility and vitality's sustenance, consonance, and safety to alleviate Charlottetown's climate-related risks. (Diagrams: designed by the author and executed by Anna Maria Levytska and Rachel Rauser)

[2] All the images in this paper are by the author unless otherwise indicated.

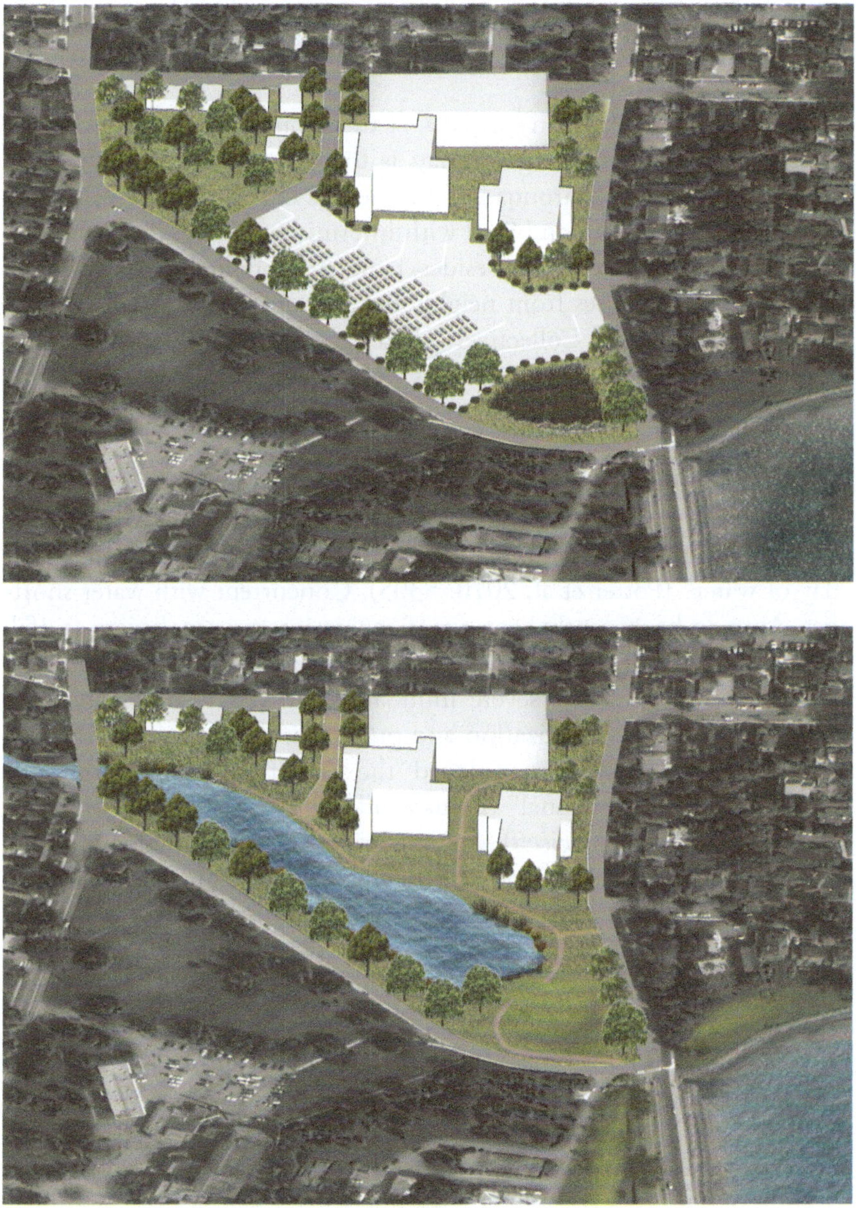

Fig. 11.1 (continued)

Khirfan and El-Shayeb forthcoming). The concept of reversibility here can help design transformative climate actions in cities where natural features have been erased.

My second case, Amman (Jordan), faces various climate stresses. Based on United Nations rankings, Jordan is the second poorest country in water in the world (Namrouqa 2014). The lack of water resources is exacerbated by severe heat combined with intermittent, yet extreme, rainfall attributed to climate change. Besides, regional geopolitics since 1948 led to an influx of refugees from neighbouring countries, who mostly concentrated in Amman.[3] Collectively, these factors led to the securitization of migration and of water resources (Weinthal et al. 2015). Amman's severe water shortages were addressed through reactive forms of climate urbanism, such as attempts to balance sustenance and adaptability through rationing household water supply to now only one day per week. Simultaneously, Amman's inhabitants adapted their behaviours by postponing water demanding household tasks and even personal needs to the "day of water" (Potter et al. 2010: 5305). Concurrent with water shortages, Amman has recently been experiencing intermittent severe rainfall especially in the downtown area. This triggered social mobilization for climate action. After the severe inundation in March 2019, and in a country where civil organization and public gatherings are heavily controlled, the citizens of Amman defied "the separation of public from private, political from personal" (Cruikshank 1993: 341). They used social media platforms to exercise their spatial rights by initiating a citizen-led marketing campaign ("Let's go downtown") that encouraged inhabitants to change their behaviours towards the long-shunned downtown and to support local merchants in their recovery from the flooding (Ayyoub 2019).

[3] Amman's growth is marked by the influx of refugees: Palestinians in 1948 and 1967; Lebanese in 1975–1990; the two Gulf Wars in 1990 and 2004; and, more recently, the wars in Syria, Libya, and in Yemen since 2011, 2014 and 2015 consecutively. According to 2015 census data that of Jordan's total population of 9.5 million, 2.9 million (30.6%) are not Jordanian citizens, nearly half of whom (49.7%) live in Amman, compared to only 38.6% of Jordanian citizens (Department of Statistics 2015; Ghazal 2016).

In Negril (Jamaica) the tourism industry embodies an entrepreneurial approach to climate urbanism. Businesses' mitigation interventions primarily entail solar energy installations while their adaptation strategies vary, as exemplified in different attempts to curb the severe beach erosion triggered by sea-level rise and extreme weather events (Fig. 11.2). Sandals Negril Beach Resort installed artificial reef and reinforced parts of its shoreline with rocks (Fig. 11.3) while small local hotels sandbagged the shoreline and installed impromptu wooden structures (Fig. 11.4). To address water shortages in the dry season, the Rockhouse, a local boutique hotel, installed rainwater harvesting tanks (Fig. 11.5). Negril's ordinary citizens are similarly creative in adaptability, fit, risk reduction, and mobilized their sense of place to regenerate the mangrove forests through improvisation (Fig. 11.6) and through protecting the Great Morass from development (Fig. 11.7) (Dhar and Khirfan 2016). All these initiatives stem from local ecological knowledge and contribute to the place's symbolic significance while also demonstrating social learning through experimentation (Anguelovski et al. 2016; Castán Broto et al. 2019; Shi et al. 2015; see also Yli-Pelkonen and Kohl 2005 on LEK).

Last, Zürich (Switzerland) initiated in 1986 a policy to daylight its network of buried urban streams—known as Bachkonzept (ERZ Entsorgung 2003). Stream daylighting refers to "the practice of removing streams from buried conditions and exposing them to the Earth's surface in order to directly or indirectly enhance the ecological, economic and/or socio-cultural well-being of a region and its inhabitants" (Khirfan et al. 2020). Adopted at the time to decrease the costs of sewerage treatment by separating it from rainwater runoff, this policy necessitated cooperation between municipal authorities and market actors to regulate its implementation in existing and new urban developments (Fig. 11.8) (personal interviews 2017; Conradin and Buchli 2007). This unique policy can be seen as a form of both entrepreneurial and transformative climate

Fig. 11.2 Severe beach erosion in Negril, Jamaica, triggered by sea-level rise and extreme weather events

urbanism as the pace and intensity of climate change-related extreme weather events increased over the last decades. Nowadays, this policy—which led to the daylighting of over 25 kilometres of streams—contributes to Zürich's distinctive identity and symbolic significance with spillover effects into the urban form's vitality. Indeed, Zürich's daylighted streams enhance its built environment and public spaces; benefit the quality of the water, flora, and fauna; and offer a multiplicity of benefits ranging from flood risk alleviation to recreational and educational benefits (Fig. 11.9) (ERZ Entsorgung 2003).

Fig. 11.3 Sandals Negril Beach Resort installed artificial reef and buffered parts of its shoreline with rocks

Fig. 11.4 Small hotels in Negril used sandbags and installed impromptu structures to address severe beach erosion and keep the seawater at bay

Fig. 11.4 (continued)

Fig. 11.5 Rainwater harvesting tanks at Negril's Rockhouse Hotel

Fig. 11.6 Attempts by locals to regenerate the mangroves in Orange Bay, Negril

Fig. 11.6 (continued)

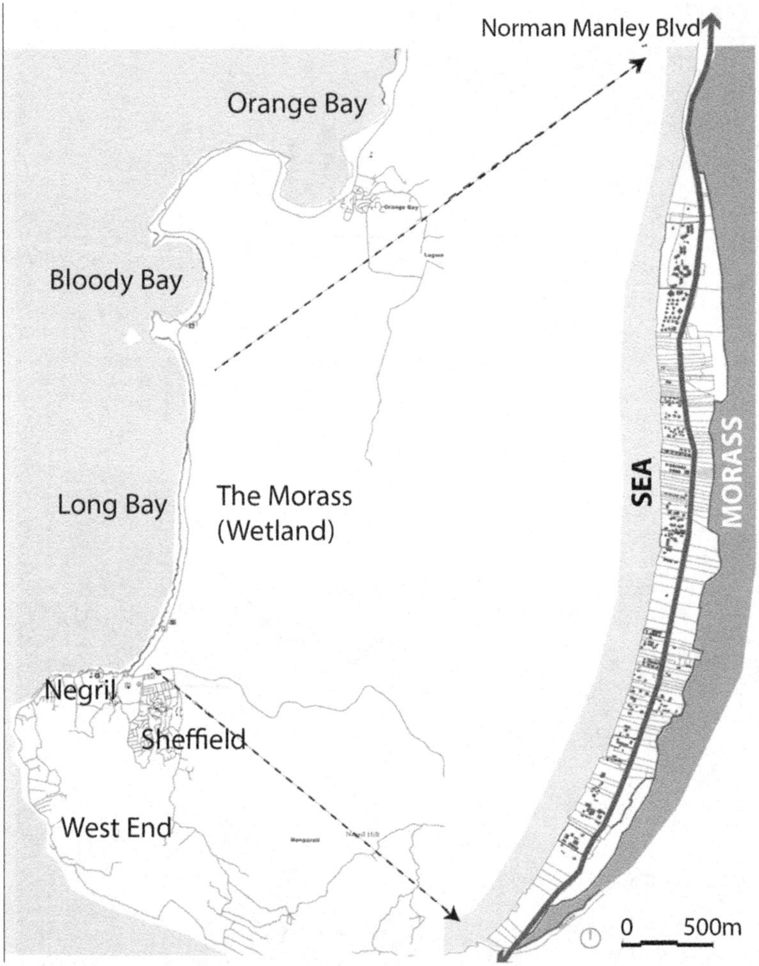

Fig. 11.7 Local activism led to the protection of Negril's Great Morass from development (Map drawn by Tapan K. Dhar)

Fig. 11.7 (continued)

Fig. 11.8 The cooperation between urban developers and municipal authorities in Zürich in implementing the Bachkonzept reflects the market's creative continuism while also transforming the urban form in unique ways

Fig. 11.9 A composite showing the positive impact of Zürich's daylighting policy as a nature-based solution that simultaneously provides vital ecosystem services and urban spaces for education, recreation, and leisure

11.4 Conclusions: Efficiency and Justice in Climate Urbanism

Lynch's overarching principles of efficiency and justice interact with all five performance dimensions and are particularly relevant for climate urbanism. Efficiency is a balancing criterion that assesses the optimization (rather than maximization) of any dimension's performance through different values and costs (including non-monetary and non-spatial ones). Justice is the way in which both monetary and non-monetary benefits and costs are spatially distributed among people and how these benefits are incorporated within all of the five performance dimensions. Such incorporation occurs through one or a combination of devices like setting minimum thresholds, equity rules on goods, and focusing on the least-favoured groups (Lynch 1981). While Lynch's spatial justice aligns more with the distributive than the procedural justice (O'Brien and Leichenko 2010), he tackles the latter through the spatial rights discussed in our analysis of control as well as through its focus on the value of a place identity (through sense) that manifest via the tangible and intangible local self-representations in the urban landscape. Empirical studies and various chapters in this book (Chaps. 3 and 8 in particular) have shown that efficiency receives more attention than justice in climate urbanism, increasing urban inequalities and rendering the urban poor vulnerable to climate-related risks (e.g. Bulkeley et al. 2013; Chu et al. 2017; Kashem et al. 2016; Keenan et al. 2018; Rice et al. 2020). Moreover, there are contradictions inherent to achieving efficiency within different dimensions of Lynch's framework. For instance, achieving efficiency in vitality for adaptation by integrating nature-based solutions into planning can contradict efforts to achieve efficient fitness for mitigation through intensification and compact urban forms (Haaland and Konijnendijk van den Bosch 2015). Conflicts emerge if adaptability leads to sacrificing present fit—as illustrated by the resistance of car users to multi-modal transportation systems accommodating public transit, cycling, and walking (Geels 2012). Unlike efficiency's conflicts, Lynch argues that the interactions between spatial justice and the other performance dimensions vary in their significance. Specifically, Lynch identifies what I consider to be scales of justice. For example, while equitable access is pivotal (e.g. evacuation to safe areas during extreme weather events) and a just distribution of control is critical (e.g. over land use planning and urban development), alternately, it might be possible to relegate the justice of fit to a secondary concern when the minimum thresholds of adaptability are met, protecting the most vulnerable. I use Amman's water scarcity case to elaborate on this. Considering global thresholds of 500 m^3 of water per capita per year, Amman's 150 m^3 threshold per capita per year is unjust (see Klassert et al. 2015). However, by providing every household in Amman with equal shares of this exact minimum baseline of water, distributed in equal durations and supply across all neighbourhoods regardless of socio-economic composition, the city is achieving a minimum threshold of fit for domestic water consumption given the challenges it faces.

Lynch's theory of Good City Form warrants further exploration for climate urbanism research not least because its five + two performance dimensions balance urban form with social and human behaviours, thus bridging existing discourses focusing on technologies as "urban structures" and social learning processes as "the drivers of change" (McCormick et al. 2013: 4–5). Considering fit and vitality, access and control, sense, and efficiency and justice enables us to tackle local particularisms in the complexity of urbanism while allowing comparisons across contexts. Most of the discourse and action on urban form and climate change are reactive and partial, and the examples presented in this chapter highlight the need for further research to explore how urban design, through modifications in the urban form, can support transformative approaches to climate urbanism.

References

Anguelovski, I., Shi, L., Chu, E., Gallagher, D., Goh, K., Lamb, Z., et al. (2016). Equity impacts of urban land use planning for climate adaptation: Critical perspectives from the global north and south. *Journal of Planning Education and Research, 36*(3), 333–348.

Ayyoub, A. (2019, March 8). Amman shoppers turn out in force to support flood-hit merchants: Residents in Jordan's capital hit the stores to back owners who suffered major losses after heavy rains led to floods. Al Jazeera. Retrieved from https://www.aljazeera.com/news/2019/03/shoppers-turn-force-support-amman-flood-hit-merchants-190308181609264.html.

Bulkeley, H., Carmin, J., Castán Broto, V., Edwards, G. A. S., & Fuller, S. (2013). Climate justice and global cities: Mapping the emerging discourses. *Global Environmental Change, 23*, 914–925.

Burch, S., Shaw, A., Dale, A., & Robinson, J. (2014). Triggering transformative change: A development path approach to climate change response in communities. *Climate Policy, 14*(4), 467–487.

Calthorpe, P. (1993). *The next American metropolis: Ecology, community, and the American dream.* New York: Princeton Architectural Press.

Castán Broto, V., Trencher, G., Iwaszuk, E., & Westman, L. (2019). Transformative capacity and local action for urban sustainability. *AMBIO, 48*, 449–462.
Chu, E., Anguelovski, I., & Roberts, D. (2017). Climate adaptation as strategic urbanism: Assessing opportunities and uncertainties for equity and inclusive development in cities. *Cities, 60*, 378–387.
Conradin, F., & Buchli, R. (2007). The Zurich stream daylighting program. In R. L. France (Ed.), *Handbook of regenerative landscape design* (pp. 47–60). Boca Raton: CRC Press.
Cruikshank, B. (1993). Revolutions within: Self-government and self-esteem. *Economy and Society, 22*(3), 327–344.
Department of Statistics. (2015). Population and Housing Census 2015. Retrieved from http://www.dos.gov.jo/dos_home_e/main/population/census2015/index.htm.
Dhar, T. K., & Khirfan, L. (2016). Community-based adaptation through ecological design: Lessons from Negril, Jamaica. *Journal of Urban Design, 20*, 1–22.
Ellin, N. (1999). *Postmodern urbanism*. New York: Princeton University Press.
ERZ Entsorgung. (2003). Streams in the city of Zürich: Concept, experiences and examples. Retrieved from Zürich, Switzerland.
Geels, F. W. (2012). A socio-technical analysis of low-carbon transitions: Introducing the multi-level perspective into transport studies. *Journal of Transport Geography, 24*, 471–482.
Ghazal, M. (2016, January 30). Population stands at around 9.5 million, including 2.9 million guests. The Jordan Times.
Haaland, C., & Konijnendijk van den Bosch, C. (2015). Challenges and strategies for urban green-space planning in cities undergoing densification: A review. *Urban Forestry and Urban Greening, 14*, 760–771.
Jivén, G., & Larkham, P. J. (2003). Sense of place, authenticity and character: A commentary. *Journal of Urban Design, 8*(1), 67–81.
Kashem, S. B., Wilson, B., & Van Zandt, S. (2016). Planning for climate adaptation: Evaluating the changing patterns of social vulnerability and adaptation challenges in three coastal cities. *Journal of Planning Education and Research, 36*(3), 304–318.
Kates, R. W., Travis, W. R., & Wilbanks, T. J. (2012). Transformational adaptation when incremental adaptations to climate change are insufficient. *Proceedings of the National Academy of Sciences of the United States of America, 109*(19), 7156–7161.

Keenan, J. M., Hill, T., & Gumber, A. (2018). Climate gentrification: From theory to empiricism in Miami-Dade County, Florida. *Environmental Research Letters, 13*(5).

Khirfan, L., & El-Shayeb, H. (2019). Urban climate resilience through socio-ecological planning: A case study in Charlottetown, Prince Edward Island. *Journal of Urbanism: International Research on Placemaking and Urban Sustainability, 13*, 187.

Khirfan, L., & El-Shayeb, H. (forthcoming). Charlottetown's climate adaptation: Reclaiming land 'from' or 'for' water? In F. Masoud & B. D. Ryan (Eds.), *Terra-Sorta-Firma: Reclaiming the littoral gradient Barcelona.* Spain: ACTAR.

Khirfan, L., et al. (2020). Digging for the truth: A combined method to analyze the literature on stream daylighting. *Sustainable Cities and Society, 59.*

Klassert, C., Sigel, K., Gawl, E., & Klauer, B. (2015). Modeling residential water consumption in Amman: The role of intermittency, storage, and pricing for piped and tanker water. *Water, 7*, 3643–3670.

Long, J., & Rice, J. L. (2019). From sustainable urbanism to climate urbanism. *Urban Studies, 56*(5), 992–1008.

Lynch, K. (1981). *Good city form* (9th ed.). Cambridge, MA: The MIT Press.

Mach, K. J., Planton, S., & von Stechow, C. (2014). Annex II: Glossary In The Core Writing Team, R. K. Pachauri, and L. A. Meyer (Eds.), Climate Change 2014: Synthesis Report. Contribution of Working Groups I, II and III to the Fifth Assessment Report of the Intergovernmental Panel on Climate Change, 117–130.

McCormick, K., Anderberg, S., Coenen, L., & Neij, L. (2013). Advancing sustainable urban transformation. *Journal of Cleaner Production, 50*, 1–11.

McHarg, I. L. (1992). *Design with nature.* New York: John Wiley and Sons.

Namrouqa, H. (2014). Jordan world's second water-poorest country. The Jordan Times. Retrieved from http://www.jordantimes.com/news/local/jordan-world's-second-water-poorest-country.

Newman, P. (2016). Emergent urbanism as the Transformative force in saving the planet. In T. Hass & K. Olsson (Eds.), *Emergent urbanism: Urban planning and design in times of structural and systematic change* (pp. 121–131). New York: Routledge.

Norberg-Schulz, C. (1980). *Genius loci: Towards a phenomenology of architecture.* NYC, Rizzoli.

O'Brien, K. L., & Leichenko, R. M. (2010). Global environmental change, equity, and human security. In R. A. Matthew, J. Barnett, B. McDonald, &

K. O'Brien (Eds.), *Global Environmental Change and Human Security* (pp. 157–176). Cambridge, MA: MIT Press.

Pelling, M., O'Brien, K., & Matyas, D. (2015). Adaptation and transformation. *Climatic Change, 133*(1), 113–127.

Pinkham, R. (2000). *Daylighting: New life for buried streams.* Colorado: Rocky Mountain Institute.

Potter, R. B., Darmame, K., & Nortcliff, S. (2010). Issues of water supply and contemporary urban society: The case of Greater Amman, Jordan. *Philosophical Transactions the Royal Society A, 368,* 5299–5313.

Rice, J. L., Cohen, D. A., Long, J., & Jurjevich, J. R. (2020). Contradictions of the climate-friendly city: New perspectives on eco-gentrification and housing justice. *International Journal of Urban and Regional Research, 44*(1), 145–165.

Shi, L., Chu, E., & Debats, J. (2015). Explaining progress in climate adaptation planning across 156 U.S. municipalities. *Journal of the American Planning Association, 81*(3), 191–202.

Wagner, M., Merson, J., & Wentz, E. A. (2016). Design with nature: Key lessons from McHarg's intrinsic suitability in the wake of hurricane sandy. *Landscape and Urban Planning, 155,* 33–46.

Weinthal, E., Zawahri, N., & Sowers, J. (2015). Securitizing water, climate, and migration in Israel, Jordan, and Syria. *International Environmental Agreements, 15,* 293–307.

Wirth, L. (1938). Urbanism as a Way of Life. *American Journal of Sociology, 44,* 1–24.

Yli-Pelkonen, V., & Kohl, J. (2005). The role of local ecological knowledge in sustainable urban planning: Perspectives from Finland. *Sustainability: Science, Practice, and Policy, 1*(1), 3–14.

12

Collaborative Education as a 'New (Urban) Civil Politics of Climate Change'

Andrew P. Kythreotis and Theresa G. Mercer

12.1 Introduction

When considering human-induced environmental changes, there needs to be more attention given to public dialogue and citizen incorporation into policy decisions, in addition to state policy and market interventions (Jordan et al. 2015; Kythreotis et al. 2019a). This signposts a need for a shift away from purely techno-managerial solutions to climate change

A. P. Kythreotis (✉)
School of Geography and Lincoln Centre for Water and Planetary Health, University of Lincoln, Lincoln, UK

Tyndall Centre for Climate Change Research, University of East Anglia, Norwich, UK
e-mail: AKythreotis@lincoln.ac.uk

T. G. Mercer
School of Geography and Lincoln Centre for Water and Planetary Health, University of Lincoln, Lincoln, UK
e-mail: TMercer@lincoln.ac.uk

V. Castán Broto et al. (eds.), *Climate Urbanism*,
https://doi.org/10.1007/978-3-030-53386-1_12

that are framed around post-political discourses (Bäckstrand et al. 2017; Kythreotis 2012a; Swyngedouw 2013) and complex international climate regime fixity (Abbott 2012; Keohane and Victor 2011) to targeting individual behaviours and emotions that can engender local public engagement from the 'bottom-up' (Chapman et al. 2017; Whitmarsh and Corner 2017). Yet, the geopolitics of the state has dominated climate and environmental policy and planning discourses at the international scale (Purdon 2015), arguably creating 'urban dystopias' whereby the local and urban scales are considered merely a reflection of international policy imperatives that create a 'short-circuiting' of knowledge flows from the urban to the international scale (Kythreotis 2018). However, there is a well-established critique citing the failure of the international climate regime (under the UNFCCC system) to tackle climate change mitigation (e.g. Harris 2013; Kythreotis 2012a, 2015; Maltais 2014; Prins and Rayner 2007; Victor 2006), alongside an emerging body of literature on climate governance *processes* operating at urban and local scales around 'experimental' governance (e.g. Bulkeley and Castán Broto 2013; Evans 2011; Evans et al. 2016; Hajer and Versteeg 2019; Hölscher et al. 2018; Kivimaa et al. 2017).

Such literatures signpost changing geographies of climate change in which the urban and local scales are increasingly playing an influential role in climate policy particularly around engaging citizens more in local and urban climate politics as a *governance* foil to the apparent failure of the *governing* international climate regime. This is also reflected in recent calls for urban scholars to identify the key research challenges that will make a difference to urban climate policy and practice going forward (van der Heijden 2019; Wolfram et al. 2019). Could we be witnessing the emergence of the international scale as a dystopic climate policy space and the urban scale as a panacea for more effective climate policy and citizen action? This certainly redresses the scalar politics of climate change whereby the local and the urban potentially subjugates the international scale in climate policy creation. Whilst urban experiments have largely been confined to governance arenas that include state and non-state actors in formal state governance settings, there is strangely a paucity of work that directly links education in schools and universities with how urban policy decision-making on climate change. In this chapter we argue that collaborative education can be viewed as a 'new (urban) civil politics' of climate change which, ultimately, has significant connotations for a 'New Climate Urbanism' that has the power to potentially enervate extra-territorial post-political pressures of state-led international climate policy.

Following this introduction, the second section briefly discusses some of the social, cultural, political, economic, scientific and spatial barriers to increased citizen engagement in climate policy and how education in schools and universities has the potential to engender what we call a 'new civil politics of climate change'. We additionally illustrate how the new civil politics of climate change has emerged from formal state-led climate policymaking and how a policy window of citizen engagement has been created to further explore the role of education in schools and universities within new regimes of climate urbanism. The third section then discusses a case study exploring how universities and schools can co-produce innovative climate action through climate change education (CCE) within the context of education for sustainable development (ESD). In conclusion, we discuss the implications of collaborative education as a new form of climate urbanism and its power to challenge traditional globally framed state-led policy discourses on climate change.

12.2 Education, Citizen Engagement and the 'New Civil Politics of Climate Change'

Before it is possible to reimagine new political reconfigurations of citizen engagement in urban climate decision-making, there are several barriers to it that warrant discussion. These include a lack of knowledge, ideologies, costs, stresses of peer pressure, emotional blockage of new knowledge, discredence, perceived risk and negative values/attitudes towards the environment (Gifford 2015; Kollmuss and Agyeman 2002). Kythreotis et al. (2019a) also point to other barriers including a need to reframe the 'climate change problem' (currently policy threatens assumptions about quality of life, fairness, progress, individual freedom), conflicts of interests (e.g. climate denialism, scientists, policymakers), the fact that citizen engagement cannot be implemented as a 'one-size-fits-all' knowledge framework (e.g. knowledge domain needs to fit with people's everyday lives—we still need experts!), uneven power relationships (e.g. between science-policy-citizens) and differences across and within countries. Given these various social, economic, political and spatial barriers there remains a caveat in whether it is possible to engender bottom-up engagement when the geopolitics of the state dominate political discourses on climate change. Can the average citizen, who does not think climate change a priority, be co-opted, even convinced that their individual behaviour can catalyse action and policy change at the local and urban scales?

There are examples of where citizens become more involved in climate research and thus indirectly, in policy decisions. Citizen Science (CS) is one way in which citizens can involve themselves at the science-policy interface by assisting scientists to obtain larger datasets to inform formal state policy practices and implementation, thus democratising expertise into more formal policy processes (Fischer 1993). All CS projects involve citizens observing and collecting data for the scientific experts, rather than them directly formulating the research methods, analysing and interpreting the data to provide evidence for climate policymaking. Hence, citizen science arguably limits full engagement and co-production to directly influence climate policy. This is despite ample evidence of urban and local governments working better to include traditional and local knowledge into their policy systems (Leonard et al. 2013; Mantyka-Pringle et al. 2017; Naess 2013).

With the above in mind, enhancing education, research and pedagogic practices could play a transformative role in climate policy and action (Perkins et al. 2018; Kythreotis et al. 2019b). Part of the solution is to try and involve all citizens to recognise the urgency of the situation of dangerous climate change, whilst bringing a sense of purpose, hope and fun in making the necessary changes to catalyse more effective bottom-up action. School education can facilitate this at an early age. The very recent School Strike for Climate around the world inspired by Greta Thunberg, the young Latinx founder of the Climate Zero action group, Jamie Margolin, and the co-founder of the This Is Zero Hour movement, Nadia

Nazar, have demonstrated how younger generations with different backgrounds could be key in getting the message across to those (adult) groups in society that have greater institutional power—and even legitimacy—in catalysing the policy changes needed to foster greater bottom-up climate action at the urban and local scales. Even more so, these young activists inspire activism through their quotidian experiences of local and urban climate struggles, illustrating the importance of a new civil politics of climate change emerging beyond formal state-led climate policy arenas like the international policy scale.

Much of the urban climate policy discourse has traditionally revolved around the politics of the state, in particular the relationship between local and national governments in the cascading of policy decisions downwards from international to local scales (e.g. Betsill and Bulkeley 2007; Bulkeley and Betsill 2013). National climate policy blockages can catalyse local policy innovation and create new configurations of polycentric governance (Gillard et al. 2017), although more recent research in the context of climate adaptation suggests that local climate governance is often dominated by national sector-led priorities (Kythreotis et al. 2020). This illustrates how the politics of climate change is often reduced to formal state policymaking by local/urban governments that incorporate particular non-state governance actors legitimised by the state (Kythreotis 2010). This tends to confine a definition of (climate) politics to the art of government. Hence, formal state-led urban climate policy can be perceived as a policy space that evacuates more amorphous interpretations of the political based on how citizens and various civil groups construct ideas and arguments related to climate change on different philosophies, ideologies and cultures. Yet, thinking of a 'climate urbanism' in such a homogeneous way will inevitably catalyse less uniform politics of action amongst different civil groups—what we call a 'new civil politics of climate change'. This is because formal state policies are often perceived by civil actors as redistributing wealth or income that in turn create salient social, political and/or cultural (in)justice issues at urban and local scales—explaining why they often act in opposition or protest to their government. The resultant urban politics of climate change will be highly conflictual—as witnessed through the current School Strikes, Indigenous struggles over Tar Sands and gas pipelines (e.g. see

Mantyka-Pringle et al. 2015) and mass disobedience marches by Extinction Rebellion (XR).

The emergence of this new civil politics of climate change has forced different levels of government to rethink how policy is created and implemented, as the declarations of climate emergencies during 2019 by all the UK devolved sub-national territories of Scotland, Wales, and Northern Ireland demonstrate. As a result, cities in the UK and beyond have also declared climate emergencies, opening up urban policy windows to enable greater citizen-led engagement in formal policy change. Firstly, and indirectly, policy windows have been created by the state by ostensibly confining the politics of climate change at the urban scale to more formal governance arenas that are local government-led. This is where pragmatic governance (e.g. local government involving the usual civil society suspects) has trumped autonomous governance (e.g. the actions of those civil groups who reject state input into their environmental objectives) (Kythreotis 2012b). Secondly, and more directly, as a result of excluding many civil actors from formal urban climate governance processes, there is a civil 'backlash' to the way in which state climate politics are conducted culminating in a civil saturation point whereby the state can no longer avoid civil calls for greater policy action and change as evidenced in this new civil politics of climate change. The next section considers the role of collaborative education in schools and universities in this new civil politics of climate change, discussing its implications for climate urbanism.

12.3 Collaborative Education for Climate Change Education

The School Strikes that have taken place in many of the world's major cities has illustrated the collective power that young people can have in communicating their views and pushing for policy change from those in power. Younger generations understand that they are likely to bear the brunt of climate change impacts in their lifetime and as such tend to have higher levels of climate change concern than older generations (Corner et al. 2015). However, for young people to be able to act and make a difference, they need to have a clear understanding of what climate change is, what the potential impacts are and what action can be taken to deal with the impacts. This is where formal education plays an increasingly crucial role in the new urban civil politics of climate change. Formal education is one appropriate channel to deliver this by providing more scientifically informed, evidence-based understandings of climate change. We are cognizant that there are also other information channels that young people can draw on and engage with (e.g. social media), however these can be tainted with climate misinformation because of certain economic and political (e.g. ideological) vested interests that often lack scientific rigour.

Mainstreaming climate change education (CCE) in the context of sustainable development through the entire education system has multiple benefits in terms of enhancing bottom-up collaborative approaches and through intergenerational learning. When young people receive effective CCE throughout their schooling experience, they are more likely to educate and influence other older members of their families, wider society and future generations (Lawson et al. 2019; Mochizuki and Bryan 2015). This results in an increase in local and urban capacity and solutions for dealing with the impacts of climate change by empowering the general public to engage in civil society and debates as well as through influencing local policy/decision-makers to make changes (Mochizuki and Bryan 2015).

Some of the issues with delivering 'effective' CCE to young people can result from different levels of uptake into the curriculum and different levels of integration. There are also issues around teaching capacity, related to lack of knowledge and dealing with a controversial value-laden topic (Gayford 2002; Johnson et al. 2008). Recent examples of where more integrative CCE have been planned and/or implemented demonstrate that researchers, educators and policymakers are working towards overcoming some of these issues. For example, in November 2019, Italy's Minister for Education announced that climate change and sustainability education will be embedded into every subject area and within every grade in all public schools (Horrowitz 2019). This is expected to be rolled out by September 2020. A different example from Vare et al. (2019)

looked at developing a practical accreditation model for educators. Their work focused on education for sustainable development (ESD) in which CCE sits. As part of the framework, they created 12 competencies (such as systems knowledge, transdisciplinarity, participation, empathy and futures) and learning outcomes that are taught and assessed in order to qualify the educator to teach within this subject. These two examples show that changes are being made in terms of policy initiatives that are delivering integrative and equal CCE delivery as well as overcoming capacity issues of educators through competency training.

Recent work has also highlighted the important role of universities in participating with and improving CCE. Non-governmental organisations such as EAUC (the Alliance for Sustainability Leadership in Education) have responded to the UK government's Climate and Environment Emergency declaration by setting up a Climate Commission for Further Education (FE) and HE institutions (EAUC 2019). As part of this, 7000 universities across six continents have created and signed up to a three-point plan of action to deal with the emergency which will focus on becoming carbon neutral by 2030, providing more resources for climate change research and increasing "education on environmental and sustainability education across curricula, campus and community outreach programmes" (O'Malley 2019). Such collective action across FE and HEI demonstrates the significant influence that these institutions have due to their social position and cultural standing. When institutions join efforts behind a common cause, their standing as agents of change is amplified (Leal Filho et al. 2018a). Universities are also heralded as key agent in closing the 'action gap' on climate change (Molthan-Hill et al. 2019).

However, there are still many perceived barriers that universities face to implement CCE action, research, teaching and learning. These include a lack of funds, a lack of experts, university culture, political agendas above university level and a lack of connectivity within institutions (Leal Filho et al. 2018a, 2019). Different levels of integration of CCE in the context of ESD can also cause issues for delivering 'effective' teaching. This is evidenced through the different levels of uptake of sustainability initiatives between universities, for example, through the various sustainability leagues, green tables, environmental strategies and mainstreaming of CCE and ESD in the curriculum. Molthan-Hill et al. (2019) explored the ways that HEI are embedding CCE into their curriculum. This is highly variable between institutions and can range from taking a narrow approach such as 'piggybacking' individual sessions and modules to existing courses, or creating specialist courses on climate change up to mainstreaming CCE across the entire curriculum using transdisciplinary approaches (ibid.). HE students are also becoming increasingly intolerant and vociferous of those institutions that appear hypocritical in terms of the disconnect between what they preach, teach and research about sustainability and climate change against how universities and academics operated day-to-day (Ross 2019). Such discontents arguably have created the conditions for a new climate urbanism. In particular, the realisation that young people within urban areas should be included in the re-definition of "what cities are and ought to be in a changing climate" (Chap. 1). This will effectuate new modalities of climate urbanism to integrate new forms of knowledge production, and to recognise that young people are central to successful climate action.

It is now widely acknowledged that academics need to transform the way that CCE and ESD is taught at university level by creating innovative and collaborative pedagogies that can lead to critical self-reflection and transformative learning (Leal Filho et al. 2018a, b). This transformative and whole university approach extends to universities' 'moral authority'—the moral responsibility to engage with the wider public outside of academia on issues of climate change, especially in their own local communities. We therefore present a pedagogical case study that demonstrates a collaborative and transformative approach to CCE, where HE students can deliver effective CCE to young people in lower-level educational settings, whilst enhancing their own learning experiences (Mercer et al. 2017). Environmental literate HE students are well placed to supplement delivery of CCE to primary and secondary schools as they have the academic knowledge, skills and accessibility to the younger generations. As part of their assessment on a second year Human Impacts on the Environment module, HE students within Geography at Keele University in the UK were tasked with using their knowledge from the course and creativity to create, develop, and deliver educational games on climate change and sustainability with local primary school children. The aim was to educate on the impacts of climate change and to influence pro-environmental behaviour. This 'learning by design' approach gave students the opportunity to become educators and influencers. The students developed four highly creative games (Mercer et al. 2017). The first game taught pupils how to correctly recycle common household items with negative impacts on the environment when these were not identified correctly. The second game used cost-benefit analysis to create an Eco-house ('Build It Green'). The third game involved a timed card game where pupils had to work against the clock to divide environmental issues into groupings of problems, effects and solutions ('Sustainabilty Snap!'). The final game was a Carbon Points Board Game where players roll a dice and work round the board. There are questions and 'fact squares' about carbon emissions that must be navigated. The game works through a carbon points system with the aim of completing the game with as few carbon points as possible.

We witnessed several benefits for the HE students that delivered and created the games as well as the primary school children that played them (Mercer et al. 2017). Overall, the HE students felt that they would like to see more of this type of innovative and engaging assessment on the curriculum. They particularly enjoyed the creativity involved in the developing process of the game. Most students found an improvement to their skills in areas such as effectively communicating science, team working, problem solving and coming up with creative solutions. Some of the students that were particularly inspired by the activity began volunteering at the Sustainability Hub at Keele University, demonstrating a 'ripple effect' that this form of public engagement can have in terms of enhancing community links and students' sustainable behaviours as well as broader engagement through intergenerational learning. The games were enjoyed by the primary school children and teachers reported that the children had learned more about their impacts on the environment. Primary school children went back to their schools with plans to hold events such as an awareness assembly or were going to present the games they played with peers at school. This demonstrates the wider impact that these activities can have in the wider school community and children's social circles outside of the educational setting.

12.4 Conclusion

The educational games presented in this chapter represent many of the strategies that are deemed effective CCE interventions as outlined by Monroe et al. (2019). These include

> focusing on personally relevant and meaningful information, using active and engaging teaching methods, engaging in deliberative discussions, interacting with scientists, addressing misconceptions, and implementing school or community projects. (Ibid.: 791)

The above strategies relate to a new form of climate urbanism emerging through a new civil politics of climate change that can involve young people more centrally. In this sense, we can certainly identify the potential for a new type of expertise and mode of knowledge production that could represent an alternative urban governance in contradistinction to the state-led, post-political discourses that currently dominate the climate polity. Perhaps future research priorities should move away from the assumption that citizen engagement within formal urban/local climate policy spaces must be reduced to being led by non-state community groups where adults play a key governance 'agitating' role towards the local state and its resulting climate policies. A future urban politics that involves a local and global politics as suggested by Cochrane (1999, 2011) may in fact find greater credence and influence through a relational and joined-up climate polity which listens to and incorporates the views of younger generations. Therefore, urban scholars should not only be thinking about how future research agendas can encompass the diversities of planetary urban conditions but should also explore the urban diversities of intergenerationality.

Acknowledgements Andrew P. Kythreotis thanks the British Academy and the Department of Business, Energy and Industrial Strategy for funding to continue this research in the context of the new civil politics of climate change (grant number SRG19\190291).

References

Abbott, K. W. (2012). The transnational regime complex for climate change. *Environment and Planning C: Government and Policy, 30*(4), 571–590.

Bäckstrand, K., Kuyper, J. W., Linnér, B.-O., & Lövbrand, E. (2017). Non-state actors in global climate governance: From Copenhagen to Paris and beyond. *Environmental Politics, 26*(4), 561–579.

Betsill, M. M., & Bulkeley, H. (2007). Looking back and thinking ahead: A decade of cities and climate change research. *Local Environment, 12*(5), 447–456.

Bulkeley, H., & Betsill, M. (2013). Revisiting the urban politics of climate change. *Environmental Politics, 22*(1), 136–154.

Bulkeley, H., & Castán Broto, V. (2013). Government by experiment? Global cities and the governing of climate change. *Transactions of the Institute of British Geographers, 38*(3), 361–375.

Chapman, D. A., Lickel, B., & Markowitz, E. M. (2017). Reassessing emotion in climate change communication. *Nature Climate Change, 7*(12), 850–852.

Cochrane, A. (1999). Redefining urban politics for the twenty-first century. In A. Jonas & D. Wilson (Eds.), *Urban growth machines: Critical perspectives two decades later. Urban public policy* (pp. 109–124). Albany, New York: State University of New York Press.

Cochrane, A. (2011). Urban politics beyond the urban. *International Journal of Urban and Regional Research, 35*(4), 862–863.

Corner, A., Roberts, O., Chiari, S., Völler, S., Mayrhuber, E. S., Mandl, S., & Monson, K. (2015). How do young people engage with climate change? The role of knowledge, values, message framing, and trusted communicators. *Wiley Interdisciplinary Reviews: Climate Change, 6*(5), 523–534.

EAUC. (2019). EAUC announces launch of new Climate Commission | EAUC. Retrieved November 14, 2019, from https://www.eauc.org.uk/eauc_announces_launch_of_new_climate_commission.

Evans, J. (2011). Resilience, ecology and adaptation in the experimental city. *Transactions of the Institute of British Geographers, 36*(2), 223–237.

Evans, J., Karvonen, A., & Raven, R. (2016). *The experimental city*. London: Routledge.

Fischer, F. (1993). Citizen participation and the democratization of policy expertise: From theoretical inquiry to practical cases. *Policy Sciences, 26*(3), 165–187.

Gayford, C. (2002). Controversial environmental issues: A case study for the professional development of science teachers. *International Journal of Science Education, 24*(11), 1191–1200.

Gifford, R. (2015). The road to climate hell. *New Scientist, 227*(3029) Reed Business Information Ltd, 28–33.

Gillard, R., Gouldson, A., Paavola, J., & Van Alstine, J. 2017. 'Can national policy blockages accelerate the development of polycentric governance? Evidence from climate change policy in the United Kingdom', Global Environmental Change, 174–182.

Hajer, M., & Versteeg, W. (2019). 'Imagining the post-fossil city: Why is it so difficult to think of new possible worlds?', Territory, Politics. *Governance, 7*(2), 122–134.

Harris, P. G. (2013). *What's wrong with climate politics and how to fix it.* Cambridge: Polity Press.

Hölscher, K., Frantzeskaki, N., & Loorbach, D. (2018). Steering transformations under climate change: Capacities for transformative climate governance and the case of Rotterdam, the Netherlands. *Regional Environmental Change, 2*, 1–15.

Horrowitz, J. (2019). Italy's students will get a lesson in climate change. Many Lessons, in Fact. – The New York Times, The New York Times, Retrieved November 14, 2019, from https://www.nytimes.com/2019/11/05/world/europe/italy-schools-climate-change.html.

Johnson, R. M., Henderson, S., Gardiner, L., Russell, R., Ward, D., Foster, S., Meymaris, K., Hatheway, B., Carbone, L., & Eastburn, T. (2008). Lessons learned through our climate change professional development program for middle and high school teachers. *Physical Geography, 29*(6), 500–511.

Jordan, A. J., Huitema, D., Hildén, M., van Asselt, H., Rayner, T. J., Schoenefeld, J., Tosun, J., Forster, J., & Boasson, E. L. (2015). Emergence of polycentric climate governance and its future prospects. *Nature Climate Change, 5*(11), 977–982.

Keohane, R. O., & Victor, D. G. (2011). The Regime Complex for Climate Change. *Perspectives on Politics, 9*(1), 7–23.

Kivimaa, P., Hildén, M., Huitema, D., Jordan, A., & Newig, J. (2017). Experiments in climate governance – A systematic review of research on energy and built environment transitions. *Journal of Cleaner Production, 169*, 17–29.

Kollmuss, A., & Agyeman, J. (2002). Mind the Gap: Why do people act environmentally and what are the barriers to pro-environmental behavior? *Environmental Education Research, 8*(3), 239–260.

Kythreotis, A. (2010). Local strategic partnerships: A panacea for voluntary interest groups to promote local environmental sustainability? The UK context. *Sustainable Development, 18*(4), 187–193.

Kythreotis, A. (2012a). Progress in global climate change politics? Reasserting national state territoriality in a "post-political" world. *Progress in Human Geography, 36*(4), 457–474.

Kythreotis, A. (2012b). Autonomous and pragmatic governance networks: Environmental leadership and strategies of local Voluntary and Community Sector groups in the UK. In D. Gallagher (Ed.), *Environmental leadership: A reference handbook* (pp. 282–294). Thousand Oaks, California: SAGE Publications.

Kythreotis, A. (2015). 'Carbon pledges: Alliances and ambitions', Nature Climate Change. *Nature Research, 5*(9), 806–807.

Kythreotis, A. (2018). Reimagining the urban as dystopic resilient spaces: Scalar materialities in climate knowledge, planning and politics. In K. Ward, A. Jonas, B. Miller, & D. Wilson (Eds.), *The Routledge Handbook on Spaces of Urban Politics* (p. 612). London: Routledge.

Kythreotis, A., Mantyka-Pringle, C., Mercer, T. G., Whitmarsh, L. E., Corner, A., Paavola, J., Chambers, C., Miller, B. A., & Castree, N. (2019a). Citizen social science for more integrative and effective climate action: A science-policy perspective, Frontiers in Environmental Science. https://doi.org/10.3389/fenvs.2019.00010.

Kythreotis, A. P., Mercer, T., Howarth, C., Jonas, A., & Castree, N. (2019b). The 'New Civil Politics of Climate Change': Can schools, colleges and universities augment government policy action? EAUC, Nov 19th, 2019. https://www.eauc.org.uk/the_new_civil_politics.

Kythreotis, A. P., Jonas, A. E. G., & Howarth, C. (2020). Locating climate adaptation in urban and regional studies. *Regional Studies, 54*(4), 576–588.

Lawson, D. F., Stevenson, K. T., Nils Peterson, M., Carrier, S. J., Strnad, R. L., & Seekamp, E. (2019). Children can foster climate change concern among their parents. *Nature Climate Change, 9*, 458–462.

Leal Filho, W., Morgan, E. A., Godoy, E. S., Azeiteiro, U. M., Bacelar-Nicolau, P., Veiga Ávila, L., Mac-Lean, C., & Hugé, J. (2018a). Implementing climate change research at universities: Barriers, potential and actions. *Journal of Cleaner Production, 170*, 269–277.

Leal Filho, W., Raath, S., Lazzarini, B., Vargas, V. R., de Souza, L., Anholon, R., Quelhas, O. L. G., Haddad, R., Klavins, M., & Orlovic, V. L. (2018b). The role of transformation in learning and education for sustainability. *Journal of Cleaner Production, 199*, 286–295.

Leal Filho, W., Shiel, C., Paço, A., Mifsud, M., Ávila, L. V., Brandli, L. L., Molthan-Hill, P., Pace, P., Azeiteiro, U. M., Vargas, V. R., & Caeiro, S. (2019). Sustainable development goals and sustainability teaching at universities: Falling behind or getting ahead of the pack? *Journal of Cleaner Production, 232*, 285–294.

Leonard, S., Parsons, M., Olawsky, K., & Kofod, F. (2013). The role of culture and traditional knowledge in climate change adaptation: Insights from East Kimberley, Australia. *Global Environmental Change, 23*(3), 623–632.

Maltais, A. (2014). Failing international climate politics and the fairness of going first. *Political Studies, 62*(3), 618–633.

Mantyka-Pringle, C. S., Jardine, T. D., Bradford, L., Bharadwaj, L., Kythreotis, A., Fresque-Baxter, J., Kelly, E., Somers, G., Doig, L. E., Jones, P., Lindenschmidt, K., & Slave River and Delta Partnership. (2017). Bridging science and traditional knowledge to assess cumulative impacts of stressors on ecosystem health. *Environment International, 102*, 125–137.

Mantyka-Pringle, C. S., Westman, C. N., Kythreotis, A., & Schindler, D. W. (2015). Honouring indigenous treaty rights for climate justice. *Nature Climate Change, 5*(9), 798–801.

Mercer, T. G., Kythreotis, A., Robinson, Z. P., Stolte, T., George, S. M., & Haywood, S. K. (2017). The use of educational game design and play in higher education to influence sustainable behaviour. *International Journal of Sustainability in Higher Education, 18*(3), 359–384.

Mochizuki, Y., & Bryan, A. (2015). Climate change education in the context of education for sustainable development: Rationale and principles. *Journal of Education for Sustainable Development, 9*(1), 4–26.

Molthan-Hill, P., Worsfold, N., Nagy, G. J., Leal Filho, W., & Mifsud, M. (2019). Climate change education for universities: A conceptual framework from an international study. *Journal of Cleaner Production, 226*, 1092–1101.

Monroe, M. C., Plate, R. R., Oxarart, A., Bowers, A., & Chaves, W. A. (2019). Identifying effective climate change education strategies: A systematic review of the research. *Environmental Education Research, 25*(6), 791–812.

Naess, L. O. (2013). The role of local knowledge in adaptation to climate change. *Wiley Interdisciplinary Reviews: Climate Change, 4*(2), 99–106.

O'Malley, B. (2019). Networks of 7,000 universities declare climate emergency. University World News. Retrieved November 14, 2019, from https://www.universityworldnews.com/post.php?story=20190710141435609.

Perkins, K. M., Munguia, N., Moure-Eraso, R., Delakowitz, B., Giannetti, B. F., Liu, G., Nurunnabi, M., Will, M., & Velazquez, L. (2018). International perspectives on the pedagogy of climate change. Retrieved November 14, 2019, from https://doi.org/10.1016/j.jclepro.2018.07.296.

Prins, G., & Rayner, S. (2007). Time to ditch Kyoto. *Nature, 449*(7165), 973–975.

Purdon, M. (2015). Advancing comparative climate change politics: Theory and method. *Global Environmental Politics, 15*(3), 1–26.

Ross, J. (2019). Universities' 'Hypocrisy' on sustainability 'Becoming untenable', Times Higher Education. Retrieved November 14, 2019, from https://www.timeshighereducation.com/news/universities-hypocrisy-sustainability-becoming-untenable.

Swyngedouw, E. (2013). The non-political politics of climate change. *ACME: An International E-Journal for Critical Geographies, 12*(1), 1–8.

Van der Heijden, J. (2019). Studying urban climate governance: Where to begin, what to look for, and how to make a meaningful contribution to scholarship and practice. *Earth System Governance, 1*, 100005.

Vare, P., Arro, G., de Hamer, A., Del Gobbo, G., de Vries, G., Farioli, F., Kadji-Beltran, C., Kangur, M., Mayer, M., Millican, R., Nijdam, C., Réti, M., & Zachariou, A. (2019). Devising a competence-based training program for educators of sustainable development: Lessons learned. *Sustainability, 11*(7), 1890.

Victor, D. G. (2006). Toward effective international cooperation on climate change: Numbers, interests and institutions. *Global Environmental Politics, 6*(3), 90.

Whitmarsh, L., & Corner, A. (2017). Tools for a new climate conversation: A mixed-methods study of language for public engagement across the political spectrum. *Global Environmental Change, 42*, 122–135.

Wolfram, M., van der Heijden, J., Juhola, S., & Patterson, J. (2019). Learning in urban climate governance: Concepts, key issues and challenges. *Journal of Environmental Policy and Planning, 21*(1), 1–15.

Part IV

Climate Urbanism as a New Communal Project

13

Community Energy Resilience for a New Climate Urbanism

Long Seng To

13.1 Introduction

Long-term climate change and increasing frequency of extreme weather events pose key challenges for the resilience of infrastructure systems particularly in cities of the Global South. At the same time, Southern countries face energy access challenges with over 840 million people without access to electricity and almost 3 billion relying on traditional fuels for cooking and heating (IEA et al. 2019). Although most people without access to sustainable energy services live in rural areas, the challenges of energy access in cities should not be overlooked. Providing electricity to 108 million urban dwellers currently without access has been challenging due to rapid urban growth, to fragile distribution networks and because the vast majority of people without energy access in cities live in informal settlements (Castán Broto et al. 2017; IEA et al. 2019). Progress towards access to clean cooking has been outpaced by population growth in both

L. Seng To (✉)
Geography and Environment, School of Social Sciences and Humanities, Loughborough University, Loughborough, UK
e-mail: L.To@lboro.ac.uk

V. Castán Broto et al. (eds.), *Climate Urbanism*,
https://doi.org/10.1007/978-3-030-53386-1_13

rural and urban areas, adding tens of millions to the global access deficit each year (IEA 2019). Cities must find new ways of meeting these energy access challenges while ensuring the resilience of energy systems to climate change.

This chapter shows that communities can play an active role in ensuring energy resilience, and argues that community energy resilience will be central to modalities of climate urbanism that are just and transformative. In what follows, I first interrogate the idea of community resilience in the context of climate urbanism. I then explore communities' energy access strategies in the face of shocks and stresses, drawing on case studies from Nepal and Malawi. The case studies were developed through (1) a study of 55 projects related to community energy resilience across the two countries identified from academic literature and project reports, and (2) a series of international workshops on the theme of 'community energy resilience and electricity systems' held in the UK (November 2018), Nepal (February 2019) and Malawi (April 2019). I conclude with reflections on the role of communities in responding to climate change in the context of a rapidly urbanising Global South. This chapter contributes to discussions on climate urbanism by examining how climate change is shaping community action and how energy is governed locally.

13.2 Community Resilience and Climate Urbanism

Communities in urban areas have been active in responding to climate change, especially in the energy sector. Community energy projects can take a variety of different forms, including demand side management (energy efficiency), renewable energy generation for both electricity and heat, and collective purchase and distribution of energy. In the Global North, examples include the transition towns movement which seeks to address peak oil and climate change through localising economic activity (North 2010) and community ownership of renewable energy generation (Braunholtz-Speight et al. 2019; van Veelen 2017). Community-owned and led sustainable energy projects that address energy access, sustainability and security together are also promoted as means to address energy challenges in the Global South (Joshi and Yenneti 2020; Marquardt and Delina 2019). These community energy projects offer opportunities for citizen participation and local income generation which are relevant in both urban and rural settings. Mapping a diversity of motivations and contexts for community energy projects, Hicks and Ison (2018) offer five measures for the strength of community orientation in a particular project including the range of actors involved, the distribution of voting power, the distribution of financial benefits, the scale of the technology and the level of community engagement.

Although local groups have presented multiple pathways for energy transitions and climate mitigation in urban areas, the concept of 'community' needs to be further interrogated (Robin and Castán Broto, forthcoming). Indeed, resources and climate risks are not equally distributed in cities: those who live in informal settlements, for example, not only have less access to energy services (Castán Broto et al. 2017) but are also less prepared for climate change as a result of poor-quality buildings and lack of critical infrastructure (Satterthwaite et al. 2020). As a result, the needs of different members of the community may be very different. Barriers to community energy resilience include differing financial resources, whereby richer households feel more able to bear the risk of technical failure, in a way that poorer households cannot. This affects the uneven uptake of innovations such as renewable energy technologies by people within the same community (McKay et al. 2007). Communities are not homogenous, and a careful mapping of communities of interest and communities of practice can build spaces for deliberative dialogue to broaden the direction of travel for energy transitions (Campbell et al. 2016). Power dynamics and hierarchies within the community, such as gender and caste, can also limit access to resources, including energy resources. Gender should be considered in community energy projects (see ENERGIA's Gender and Energy Research Programme) as it intersects with other kinds of exclusion, such as age and disability (Castán Broto and Neves Alves 2018). Communities need different capabilities to engage with energy initiatives. Energy systems literacy, project community literacy and political literacy are needed for community participation in the design and implementation of energy projects (Cloke et al.

2017). Furthermore, the boundaries of community energy are not clear cut. In practice, community energy projects bring together community, state and private sector actors operating across multiple scales (Creamer et al. 2018). Local movements can link to cosmopolitan communities of climate risk across borders (Beck et al. 2013), such as Scotland's experience in community energy projects and their historic link with Malawi, creating opportunities for community organisations interested in energy transitions to collaborate (Davis et al. 2011). In the context of climate urbanism research, it is essential to pay further attention to the roles, relationships and practices of different types of actors to ensure just transitions.

Community action is not only important for improving access to energy and mitigating climate change, but also for climate adaption and urban resilience planning. Resilience has been used as a framework for collective action that recognises uncertainty and complexity (Grove 2018). Energy has been identified as a crucial component of urban resilience, for example, by meeting basic needs that support health and well-being in cities (ARUP 2015). In the context of disaster risk reduction, the International Federation of Red Cross and Red Crescent Societies defines community resilience as the ability of individuals, communities, organizations or countries exposed to disasters, crises and underlying vulnerabilities to anticipate, prepare for, reduce the impact of, cope with and recover from the effects of shocks and stresses without compromising their long-term prospects (IFRC 2014).

This is a useful starting point for our discussion on climate urbanism because it acknowledges multiple levels of organisation and time scales involved in resilience building, and recognises people's aspirations for long-term sustainability. Figure 13.1 offers a framework to analyse community energy resilience in the context of disaster risk reduction, whereby energy infrastructure forms a part of energy resilience, community resilience and governance structures. Energy resilience requires technical innovations, strong linkages between energy infrastructure and community resilience, and adequate governance mechanisms. Community energy resilience should be about enabling access to energy services with attention to inclusion and capacity development at all levels, as opposed to simply securing energy supplies.

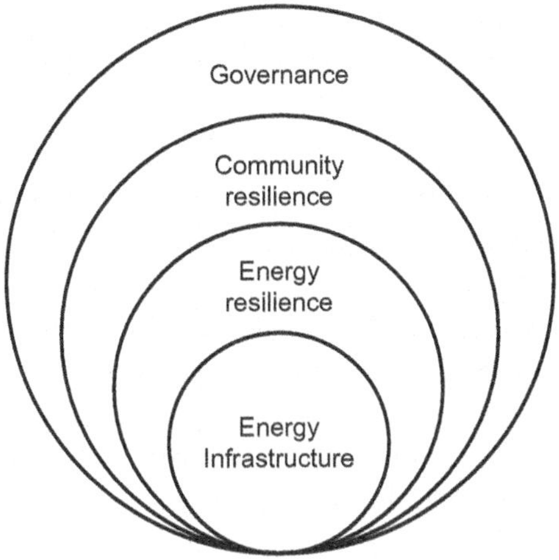

Fig. 13.1 A conceptual framework for community energy resilience (To and Subedi 2020)

Climate change presents new challenges which will increase the vulnerability of communities, within and beyond cities. Discourses on climate resilience in urban contexts have sometimes portrayed "nature as a threat to rather than an asset for cities" and cast climate change problems as a security problem (Davoudi 2014). This chapter counters this discourse by highlighting the capacity of people living in cities to cope with and to contribute to building energy resilience, opening up a vision of a socially just and transformative climate urbanism. It posits that progress towards the Sustainable Development Goals in the rapidly urbanising Global South can only be ensured by building more local resilience into critical infrastructure such as energy systems while accelerating access to energy services.

13.3 Learning from Nepal and Malawi

This section draws on examples from Nepal and Malawi to explore the challenges and opportunities of a community resilience approach to climate policy in contexts of climate stress and low levels of energy access.

Nepal and Malawi are two of the world's least urbanised (21% in Nepal and 17% in Malawi) and fastest-urbanising countries (United Nations 2018). In both countries, most of the urban population lives in informal settlements (54% in Nepal and 67% in Malawi in 2014) (Ritchie and Roser 2018). This places intense pressures on urban infrastructures, including energy, as cities continue to grow. Electricity access stands at 96% in Nepal and only 13% in Malawi (IEA et al. 2019). In addition, only 29% of the population in Nepal and 2% in Malawi have access to modern-cooking fuels (IEA 2019). In this context, the Gorkha earthquakes in Nepal (2015) and Cyclone Idai in Malawi (2019) highlighted the vulnerability of energy systems in both countries. Such extreme weather events are likely to be more frequent in the future as a result of climate change (IPCC 2014), thus posing key challenges for the resilience of infrastructure systems in both countries.

Apart from the major earthquake in 2015, Nepal is vulnerable to flooding and landslides. Climate change will exacerbate these risks and cause a loss of at least a third of the glaciers in the Hindu Kush and Himalaya range which are a critical water source (Sharma et al. 2019). Changes to water flows as a result of climate change will affect the output of hydropower stations on which Nepal relies on for of its electricity supply. These changes will affect the reliability and availability of energy services in urban areas. Nepal is also reliant on energy imports from India and a strong driver for energy resilience in Nepal is energy security (Herington and Malakar 2016). After almost a decade of negotiation following the end of the civil war, the new constitution adopted in 2015 gives greater responsibilities to local governments which consist of 6 metropolises, 11 sub-metropolises, 276 municipalities and 460 rural municipalities. This decentralisation of governance means a greater role for cities in planning for energy access as well as building resilience. This means that urban authorities have an opportunity to integrate energy into planning processes to meet increasing demands for resources and services, opening up space for new approaches to climate urbanism as the institutional arrangements evolve. Enhancing energy resilience will require careful consideration of energy access alongside measures to prevent, prepare, respond to and recover from disasters. On the technical side, quality control guidelines in systems design and installation as well as the integration of mini-grids into national grids are potential areas for increasing the reliability of energy systems. However, the benefits of resilience need to be weighed against increased costs. Energy efficiency measures may play an important role in improving resilience while keeping overall costs affordable. There is a need for coordination between operational, regional and national level actors to improve resilience. Although energy was included in Nepal's needs assessment after the Gorkha earthquake (NPC 2015), monitoring of energy during the reconstruction process did not continue. More emphasis on energy could increase opportunities to 'build back better' energy systems during reconstruction, including the ability to feed into long-term development goals through productive uses of energy for agriculture and enterprises.

Community organisations played a critical role in restoring energy services in Nepal. In the immediate aftermath of the earthquake, the National Electricity Authority found it difficult to reach some areas to restore power due to impassible roads and a shortage of gasoline fuel for transport. Community Rural Electricity Entities (CREEs) which were already operating in some rural areas as electricity distributers were able to restore supply to some households using temporary means. However, local technicians did not have the required safety tools and procedures in place (Buraityte et al. 2016). Despite these challenges, this example shows that community organisations can offer new and effective ways of responding to disasters and that providing funding and training to community organisations such as CREEs offers an alternative pathway to restoring electricity supply and build resilience going forward. Furthermore, the review shows that a small number of energy and climate resilience projects were initiated by communities to support livelihoods and to adapt to climate change. Fisherfolk living in Ramechap identified solar-powered pumps as a promising way of adaption to extreme temperatures and attaining water from underground reserves, and the community instigated the process of acquiring and installing the solar panels (Joshi 2017). In Dhye, a village decided to relocate to Thanghung due to reduced water supply and lack of arable land, which were expected to worsen due to climate change. The community chose to pursue better energy access and economic opportunities while maintaining social ties (Bernet et al. 2014; Laudari et al. 2014). Relocation and migration were part of their community energy resilience strategies. In addition, Nepal's extensive experience with implementing renewable energy projects suggest that the most successful projects tended to activate community involvement in the operation and maintenance of the energy systems by forming local groups (e.g., the UNDP Rural Energy Development Program [REDP] see Maharjan 2014; Yadoo 2012). Other important factors include additional development activities to increase livelihood opportunities and affordability of energy services and extensive testing of hardware to ensure suitability (McKay 2010; Zahnd 2014). However, renewable energy projects have tended to be externally funded and led, so it is difficult to sustain the energy access outcomes beyond the life of the project or subsidy scheme, as communities may not have the financial or technical resources to repair systems after a disaster. This threatens long-term resilience. In the review of Nepalese case studies, little overlap between renewable energy projects and resilience projects was found. Resilience projects tended to focus on ecosystems, climate and economic circumstances. Energy was rarely mentioned directly, even though energy services can provide opportunities to build resilience, for instance, supporting new income-generating activities and reducing the depletion of natural resources. The implication for climate urbanism is that although communities can be an active force in energy resilience, this predicates on existing resources and capabilities. Support for community energy resilience as a modality for climate urbanism must go hand in hand with a commitment to social justice.

Malawi has experienced cycles of drought and flooding that have led to food insecurity, including major food shortages in 2012–2013 and 2015. In March 2019, Cyclone Idai caused widespread flooding in the southern part of the country, including Blantyre (Malawi's second largest city) (ACAPS 2019a). Flooding damaged two hydroelectric powerplants, taking 270 MW out of Malawi's 320 MW of hydroelectricity offline and caused widespread disruption to electricity access in urban areas (ACAPS 2019b). Building energy resilience in the aftermath of these disasters could address these risks as well as the aspiration to improve livelihoods by selling higher value crops, the demand for reliability in agro-industry operations, reducing running costs of diesel generators, and improving availability and reliability of electricity. Direct lessons learnt from Cyclone

Idai include installing equipment to prevent debris from floods from damaging hydroelectric generators, and highlighting the need for local capacity in terms of technical design and project development. Further research is needed on the impact of disasters such as Cyclone Idai on energy systems and on the role of decentralised energy in improving energy resilience in Malawi. Indeed, the renewable energy sector is less well developed in Malawi than in Nepal. The majority of community-based solar photovoltaic projects struggle with sustaining energy access due to limited community engagement, lack of key management positions, lack of training, inability to meet operation and maintenance costs, and undersized systems which are likely to lead to technical failure (Dauenhauer et al. 2020).

However, Malawi offers interesting pathways for greater community involvement as the country is undergoing a process of decentralisation with a pilot project to create District Energy Officer roles to support local government in improving energy access (Buckland et al. 2017; Zalengera et al. 2020). Community energy projects have been developed as pilot projects to encourage district-level governments and communities to identify areas of protentional power generation. For example, the Kauvuzi Mini-hydro Power Project in Nkhatabay District will provide 50kW to about 500 households and small and medium enterprises in the area. The area has existing self-built induction machines to generate power in local streams but those are deemed unsafe. The project aims to consolidate this indigenous knowledge on small-scale hydropower generation and improve safety. The micro-grid will be owned and managed by the Kavuzi community through the creation of a new registered association (DEA 2020). The experience in Malawi shows that community involvement in the siting of renewable energy systems can be very important where flooding is a regular occurrence. Indeed, the review also found that a number of projects were ineffective because they were built in flood risk locations that were avoided by local people because of their knowledge of flood events. In some cases, systems were destroyed by flooding, in other cases, investments targeted places where people refused to live. Thus, effective resilience planning should include multiple sources of local knowledge and input to ensure the longevity of renewable energy projects and their suitability to local needs.

13.4 Conclusions

This chapter has examined the link between community resilience and climate urbanism, reviewing community projects in Nepal and Malawi to analyse the role of communities in responding to disasters and creating energy resilience. The review offers important insights into how new climate urbanisms might develop within rapidly urbanising areas in the Global South through communal responses to energy resilience. Comparing the situation in Nepal and Malawi, three important considerations for climate urbanism have emerged. Firstly, the context for resilience is important. Nepal and Malawi face different hazards and this means that the priorities for energy resilience in each context were different. Secondly, we know that the most successful energy projects include strong community participation in project design and management, but we are also learning that in the context of climate change local knowledge can help inform the careful siting of new energy systems. This highlights the need for community engagement and integration of local knowledge into urban planning processes to meet increasing energy access challenges as cities expand, particularly in informal settlements. Lastly, communities have shown that they can be active participants in responding to disasters, so response plans need to take this into account. This is not new to international development, but it is something that rapidly urbanising countries in the Global South will need to embrace if they are to develop truly resilient responses to intensified climate change.

Acknowledgements This work is supported by the Royal Academy of Engineering under the Research Fellowship scheme. The workshops on 'community energy resilience and electricity systems' were a collaboration between the Low Carbon Energy for Development Network and the Energy and Economic Growth Programme, with funding from the UK Energy Research Centre Whole Systems Networking Fund. Workshop participants included practitioners, policymakers and academics from Nepal, Bangladesh, India, Myanmar, Sri Lanka, Malawi, Kenya, Zimbabwe, Zambia, South Africa, Mozambique, USA and the UK. Special thanks to Kaitlyn Law, Vincent Mwale and Kondwani Gondwe for contributing to the database of community energy resilience projects in Nepal and Malawi, and to Nipunika Perera from the International Institute for Environment and Development for providing valuable comments on a draft of this chapter.

References

ACAPS. (2019a). Malawi Floods: Briefing note – 12 March 2019. Retrieved November 4, 2020, from https://reliefweb.int/report/malawi/malawi-floods-briefing-note-12-march-2019.

ACAPS. (2019b). Malawi Floods: Update 1 Briefing note – 19 March 2019. Retrieved November 4, 2020, from https://reliefweb.int/report/malawi/malawi-floods-update-1-briefing-note-19-march-2019.

ARUP. (2015). *City resilience framework*. London: ARUP.

Beck, U., Blok, A., Tyfield, D., & Zhang, J. Y. (2013). Cosmopolitan communities of climate risk: Conceptual and empirical suggestions for a new research agenda. *Global Networks, 13*, 1–21.

Bernet, D., Baumer, M., Devkota, F., & Pittet, S. L. (2014). Moving down or not? Phase II: Dheye, Kam for Sud.

Braunholtz-Speight, T., McLachlan, C., Mander, S., Cairns, I., Hannon, M., Hardy, J., Manderson, E., & Sharmina, M. (2019). *Visions for the future of community energy in the UK*. London: UKERC Energy Research Centre.

Buckland, H., Blanchard, R., Sieff, R., Eales, A., Yona, L., Nyirenda, E., Brown, E., Zalengera, C., Bayani, E., Cloke, J., & Batchelor, S. (2017). Malawi district energy officer blueprint: Recommendations Paper 15.

Buraityte, A., Maharjan, R., Sharma, M., & Ghimire, D. (2016). Disaster Risk Reduction Capacity of Community Rural Electricity Entities (Briefing paper No. 1), Resilience Initiative in Mechi-Mahakali-Mustang-Marchawar (RIM4). Discussion Series. Institute of Social and Environmental Transition-Nepal, Kathmandu, Nepal.

Campbell, B., Cloke, J., & Brown, E. (2016). Communities of energy. *Economic Anthropology, 3*, 133–144.

Castán Broto, V., & Neves Alves, S. (2018). Intersectionality challenges for the co-production of urban services: Notes for a theoretical and methodological agenda. *Environment and Urbanization, 30*, 367–386.

Castán Broto, V., Stevens, L., Ackom, E., Tomei, J., Parikh, P., Bisaga, I., To, L. S., Kirshner, J., & Mulugetta, Y. (2017). A research agenda for a people-centred approach to energy access in the urbanizing global south. *Nature Energy, 2*, 776–779.

Cloke, J., Mohr, A., & Brown, E. (2017). Imagining renewable energy: Towards a social energy systems approach to community renewable energy projects in the Global South. *Energy Research and Social Science, Narratives and Storytelling in Energy and Climate Change Research, 31*, 263–272.

Creamer, E., Eadson, W., van Veelen, B., Pinker, A., Tingey, M., Braunholtz-Speight, T., Markantoni, M., Foden, M., & Lacey-Barnacle, M. (2018). Community energy: Entanglements of community, state, and private sector. *Geography Compass, 12*, e12378.

Dauenhauer, P. M., Frame, D., Eales, A., Strachan, S., Galloway, S., & Buckland, H. (2020). Sustainability evaluation of community-based, solar photovoltaic projects in Malawi. *Energy, Sustainability and Society, 10*, 12.

Davis, G., MacKay, R., MacRae, M., Nicolson, L., Currie, C., MacPherson, R., Banda, E., Tembo, K., Ault, G., Frame, D. F., & Picken, S. (2011). Supporting community energy development in Malawi: A scoping study for the Scottish Government (Report). Scottish Government.

Davoudi, S. (2014). Climate change, securitisation of nature, and resilient urbanism. *Environment and Planning C: Government and Policy, 32*(2), 360–375.

DEA. (2020). Kavuzi Mini-hydro Power Project [WWW Document]. Department of Energy Affairs, Republic of Malawi. Retrieved November 11, 2020, from https://energy.gov.mw/index.php/projects/mini-grids/kavuzi.

Grove, K. (2018). *Resilience, Key ideas in geography*. London: Routledge; Taylor and Francis Group.

Herington, M. J., & Malakar, Y. (2016). Who is energy poor? Revisiting energy (in)security in the case of Nepal. *Energy Research and Social Science, 21*, 49–53.

Hicks, J., & Ison, N. (2018). An exploration of the boundaries of 'community' in community renewable energy projects: Navigating between motivations and context. *Energy Policy, 113*, 523–534.

IEA, IRENA, UNSD, WB, WHO. (2019). 2019 Tracking SDG7: The Energy Progress Report. Washington, DC.

IFRC. (2014). *IFRC Framework for Community Resilience*. Geneva: International Federation of Red Cross and Red Crescent Societies.

IPCC. (2014). Climate Change 2014: Synthesis Report. Contribution of Working Groups I, II and III to the Fifth Assessment Report of the Intergovernmental Panel on Climate Change. IPCC, Geneva.

Joshi, A. R. (2017). Solar power, sprinkle irrigation enrich vulnerable Nepali village. The Third Pole. URL. Retrieved April 17, 2020, from https://www.thethirdpole.net/en/2017/10/11/solar-power-drip-irrigation-enrich-vulnerable-nepali-village/.

Joshi, G., & Yenneti, K. (2020). Community solar energy initiatives in India: A pathway for addressing energy poverty and sustainability? *Energy and Buildings, 210*, 109736. https://doi.org/10.1016/j.enbuild.2019.109736.

Laudari, R., Kandel, B. R., & Baidya, S. L. (2014). Renewable energy options to strengthen climate resilience of mountain communities, 36–39.

Maharjan, S. (2014). Technical problem analysis of micro hydro plants: A case study at Pokhari Chauri of Kavre District. *Journal of the Institute of Engineering, 10*, 149–156.

Marquardt, J., & Delina, L. L. (2019). Reimagining energy futures: Contributions from community sustainable energy transitions in Thailand and the Philippines. *Energy Research and Social Science, 49*, 91–102.

McKay, K. H. (2010). Socio-cultural dimensions of cluster vs. single home photovoltaic solar energy systems in rural Nepal. *Sustainability, 2*, 494–504.

McKay, K. H., Zahnd, A., Sanders, C., & Nepali, G. (2007). Responses to innovation in an insecure environment in rural Nepal. *Mountain Research and Development, 27*, 302–307.

North, P. (2010). Eco-localisation as a progressive response to peak oil and climate change – A sympathetic critique. *Geoforum, Themed Issue: Geographies of Peak Oil, 41*, 585–594.

NPC. (2015). *Nepal earthquake 2015 post disaster needs assessment.* Nepal: National Planning Commission, Government of Nepal, Kathmandu.

Ritchie, H., & Roser, M. (2018). Urbanization. Our world in data. Retrieved from https://ourworldindata.org/urbanization.

Robin, E., & Castá Broto, V. (forthcoming). Intersectionality aspects of community energy: Challenges and conflict resolution methodologies. In A. Rigon & V. Castán Broto (Eds.), *Inclusive urban development in the global south: Intersectionality, inequalities, and community.* London: Routledge.

Satterthwaite, D., Archer, D., Colenbrander, S., Dodman, D., Hardoy, J., Mitlin, D., & Patel, S. (2020). Building resilience to climate change in informal settlements. *One Earth, 2*, 143–156.

Sharma, E., Molden, D., Rahman, A., Khatiwada, Y. R., Zhang, L., Singh, S. P., Yao, T., & Wester, P. (2019). Introduction to the Hindu Kush Himalaya assessment. In P. Wester, A. Mishra, A. Mukherji, & A. B. Shrestha (Eds.), *The Hindu Kush Himalaya assessment: Mountains, climate change, sustainability and people* (pp. 1–16). Cham: Springer International Publishing.

To, L. S., & Subedi, N. (2020). Towards community energy resilience. In O. Grafham (Ed.), *Energy access and forced migration* (pp. 83–93). London: Routledge.

United Nations. (2018). World Urbanization Prospects: The 2018 revision. United Nations, Department of Economic and Social Affairs, Population Division.

van Veelen, B. (2017). Making sense of the Scottish community energy sector – an organising typology. *Scottish Geographical Journal, 133*, 1–20. https://doi.org/10.1080/14702541.2016.1210820.

Yadoo, C. (2012). The role for low carbon electrification technologies in poverty reduction and climate change strategies: A focus on renewable energy mini-grids with case studies in Nepal, Peru and Kenya. *Energy Policy, 42*, 591–602.

Zahnd, A. (2014). The role of renewable energy technology in holistic community development. Springer Science and Business Media. https://doi.org/10.1007/978-3-319-03989-3.

Zalengera, C., To, L. S., Sieff, R., Mohr, A., Eales, A., Cloke, J., Buckland, H., Brown, E., Blanchard, R., & Batchelor, S. (2020). Decentralisation: The key to accelerating access to distributed energy services in sub-Saharan Africa? Journal of Environmental Studies and Sciences.

14

Making Climate Urbanism from the Grassroots: Eco-communities, Experiments and Divergent Temporalities

Jenny Pickerill

14.1 Introduction

Eco-communities are forms of grassroots transformation: they are hopeful spaces of resident-led collective change (Cattaneo 2015). Eco-communities are crucial elements of the new climate urbanism as key sites for action for adaptation and mitigation efforts, be that as spaces of innovative and inspiring climate-proofing or as exclusionary eco-enclaves (Rice et al. 2020; Silver 2018; Hodson and Marvin 2010; Anderson 2017). Eco-communities have their limitations but examining them helps us understand how urbanism can and will evolve in a climate-changed world, even if this transformation might be partial, uneven, and requires significant time. Even as apparently defensive spaces of retreat, eco-communities remain grassroots and offer holistic responses to climate change that seek to transform all elements of how we live. Therefore, they are important sites in re-defining what cities ought to, or could, be

J. Pickerill (✉)
Department of Geography, University of Sheffield, Sheffield, UK
e-mail: j.m.pickerill@sheffield.ac.uk

V. Castán Broto et al. (eds.), *Climate Urbanism*,
https://doi.org/10.1007/978-3-030-53386-1_14

in a changing climate. Despite a concentration in rural spaces (driven largely by cheaper land costs), in recent years eco-communities have purposefully sought out urban locations in order to respond to the particular pressures of housing availability, carbon-intensive transport, and sustainable livelihoods.

This chapter draws upon my ongoing research into how eco-communities are built, operate, maintain themselves and grow, along with the variety of challenges they face, to explore what we can learn from them when examining possibilities for climate urbanism. Eco-communities build upon and produce particular types of knowledge that are often disconnected from urban and policy governance approaches but are crucial in understanding the transformational potential of grassroots actions. I use the concept of infrastructures—in their broad and multi-faceted form—as a potential way to bridge the intellectual, and sometimes physical, distance between eco-communities and debates about climate urbanism. Through experimental and self-built physical and social infrastructures, eco-communities have self-provided for their basic needs—water, energy, sanitation, food, education, homes and livelihoods. Devising low-tech, low-budget, low-skill systems, which are partially off-grid and on the edges of neoliberalism, these infrastructures seek to counter the smart city, technological and top-down urban approaches. Therefore, I am asking the question: what can cities and climate urbanism research learn from the multiple types of self-built infrastructures of eco-communities in the quest to create climate change proof living spaces?

14.2 Conceptual Starting Points

In exploring what eco-communities offer to climate urbanism, there are four conceptual starting points that can help us understand how grassroots transformations occur. Eco-communities tend to start from a quest of self-provision—a grounded set of practices to detach from centralized and State provision (rather than any plan to change, for example, urban governance processes). To self-provide, they engage in ongoing forms of experimentation, through which they design and build their own (physical and social) infrastructures, all of which take considerable time. This context is important in examining urban transformations because conceptually eco-communities deliberately start small and incrementally, in contrast to other forms of climate urbanism such as metropolitan climate adaptation plans, large-scale infrastructure retrofitting, discussed in this book. Thus, the first important conceptual starting point is the idea of *self-provision*. I start exploring this question from an interest in grassroots, DIY, often anarchist, prefigurative politics (Pickerill and Chatterton 2006)—a belief that we can have agency over our lives and our worlds, albeit to different extents depending on privilege and place. Eco-communities are great examples of self-provision, building their own homes (Pickerill 2016), developing collective sustainable practices (Ergas 2010), and sharing infrastructures, responsibilities, and material objects (Jarvis 2019). The second conceptual starting point is the notion of *experiments*. Experimentations are not necessarily temporary nor spatially bounded; experiments overflow into their surroundings, creating "complex cartographies" (Davies 2010). While many experiments involve material changes, such as developing new infrastructures for growing and sharing food, they are social as much as material, often requiring changes to social practices, relations and expectations. Change often needs to be collective, a process of encouraging each other in shifting cultural expectations of what is acceptable. Understanding these social dynamics, particularly how people reconfigure what they deem acceptable and appropriate, is central to enabling transformative change (Pickerill 2019; Fois 2019). Even seemingly politically progressive experimentations can be built on troubling exclusions. What might initially appear as alternative forms of transformation can be built on neoliberal rationalities, reproducing neoliberal conditions that undermine their radical potential. Argüelles et al. (2017) summarize such rationalities as a focus on individual responsibility rather than calling for State intervention, which in turn "might help to legitimize neoliberal attempts of disposing the State from its economic and societal functions" (ibid.: 38). This ability to retreat from the State is reliant on the "privileged progressive whiteness that permeate" (ibid.: 40) these experiments, an environmental and social privilege that enables such individuals to self-provide, self-organize and improve their quality of life. The absence of a critical analysis of privilege and power in such experiments means the broader political possibilities of transformative change are limited. There is a need to be vigilant to the *politics* of experiments (Powell and Vasudevan 2007) and their *justice* (Caprotti and Cowley 2017). What is powerful about experimentations is their messy, unfinished, fluid and open nature, their unbounded, overflowing implications, but only if we also engage in what is not achieved, undone or remade. The third conceptual starting point is that of *infrastructures*. I am particularly interested in the interrelationships between physical and social infrastructural changes, the material and organizational structures of social life, especially those forms built through self-provision and owned collectively. These informal experimental infrastructures mirror elsewhere as heterogeneous infrastructure configurations (Lawhon et al. 2018), where informality and necessity combine to create space and potential for "radical incrementalism" (Pieterse 2008) and where infrastructures are reconfigured by urban dwellers to address issues with service provision and network failures. Crucially, the informality of these infrastructures, just as in eco-communities, can be liberatory but also oppressive because they can appear incoherent, they are easily disrupted and prone to failure and often small scale. The final and fourth conceptual starting point requires us to pay attention to *temporalities*. Woven into these questions about self-provision, experiments and infrastructures are questions about temporalities. In understanding temporalities, there are three aspects of time that are particularly relevant and useful. First, our experience of time is shaped by our relationship to the future. Human projects are propelled towards the future. We live, in large part, in anticipation of this future. Second, time is unstable, continuously being made and non-linear. A small change in, for example, river flow speed, can become amplified to trigger a response out of proportion, a 'jump'. The whole character of an ecosystem can change if a threshold is crossed. Therefore, vulnerability to major changes depends on where this threshold is. Third, nature's temporalities, and how environmentalists interpret these, shape their understanding of time and change. Plants, seasons, light and climate all create and shape time and humans can learn to be affected by these, or choose to live according to what Pascoe (2014) calls 'agricultural time'—that is, a time of action fit to the changing rhythms of nature. However, nature does not have a stable equilibrium state or a moment where it is perfectly balanced, rather it has multiple stable states which nature alternates, or oscillates, between (Holling and Gunderson 2002). These four conceptual starting points illuminate the need to explore the possibilities of transformation in climate urbanism through questions of who makes changes, who benefits, in what timescale, and with what success and permanency.

14.3 Eco-communities

Eco-communities are collective and collaborative projects that seek to balance human and environmental needs (Pickerill 2016). It is deliberately defined here as a broad concept that encompasses eco-villages, intentional communities, cohousing and low impact developments, among others (Pickerill 2015b). Eco-communities are ecological and community orientated; they build, make and enact new socio-materialities. In the context of climate urbanism, it is the holistic approach taken by eco-communities that recognize the interdependencies between, for example, housing and livelihood, in generating climate change and, therefore, in how cities need to respond, which yields their transformative potential. Their practices can be categorized into five main overlapping activities, which is the self-provision of: (1) homes; (2) livelihoods; (3) physical systems; (4) production; and (5) education. Eco-communities are interesting on multiple levels, as they are:

- *Active:* They are dynamic, in constant flux, spaces of doing, making and creating. Building new infrastructures of energy provision, sewerage, and so on;
- *Experimental:* They are spaces of invention and innovation—spaces of experiments, experimental spaces;
- *Environmental:* They seek to reduce their environmental impact in novel ways;
- *Collective:* They are self-organized and collaborative. They develop innovative decision-making systems;
- *Anti-capitalist:* They attempt to disrupt capitalist hegemony and anti-capitalist practices by operating beyond capitalist relations.

Eco-communities often encourage home- or land-based livelihoods and minimize economic needs, some share income;
- *Future-orientated:* They are invested in constructing alternative futures.

But eco-communities also have their own limitations. They often *lack diversity and accessibility*. They represent a narrow demographic of the population—often highly educated, white, able-bodied and with a greater proportion of women (Chitewere 2018; Bhakta and Pickerill 2016). There is an expectation that residents need to be physically fit and emotionally resilient. They tend to be *homogenous*. While eco-communities may appear aesthetically different from conventional norms, there is an environmentalist aesthetic which is common between eco-communities, even on a global scale. In other words, an eco-community kitchen in Thailand looks much the same as one in Spain. In addition, eco-communities are *slow and hard*, in that they tend to deliver slow change and require hard work from those living in eco-communities. Eco-communities are also *tied to capitalism and the State*, as they can replicate, repeat and mirror conventional society in multiple ways (gender relations, the way money is used, etc.), and rely on state support. Eco-communities have elements that embrace conventional neoliberal values. However, I am interested in the hope and promise of the everyday practices and innovations that happen in eco-communities and in how those disrupt and transgress conventional society, particularly as components for building climate-resilient and just futures.

14.4 Self-built Infrastructures

I have been working with a range of eco-communities in Britain, Spain, Thailand, USA, Australia and Argentina since 2005, focusing mostly on the Global North. Most recently, I conducted research with 12 eco-communities in Britain in the summer of 2016, which is what this section draws on. My methods include interviews, observation, and photography. Using two examples of practices of self-provisioning from my case studies, it is possible to examine how eco-communities can deliver alternatives forms of infrastructure.

Reinventing Physical Infrastructures: The Case of Greenhills (UK)

Green Hills is an eco-community that is entirely off-grid (Fig. 14.1). Will and May rely on a wind turbine and solar panels for electricity and bottled gas and a wood stove for heating water and cooking. Rainwater is pumped into their house, but spring drinking water has to be collected from a tap down the hill. They have gradually built the infrastructure themselves, over the years:

> *Every now and again we have one of those little landmark moments like 'oh that tap's suddenly been put in' or 'that pipe has been put in'. So getting the water from A to B is suddenly a lot easier. I think about when we were first here and water had to be brought in from offsite, because there's no mains water here and, now we've got a well with a pump that takes the water to inside to a sink in Matt and Jo's house and to a tap outside the front of our house. It's only a matter of time before I put a pump on our house that'll bring that water inside our house as well.* (Will, Green Hills, interview)

Fig. 14.1 Green Hills eco-community

They do not have a bathroom, but they do have a compost toilet and separate urinal spots in the woods. The infrastructure Green Hills members have self-built forces them into certain environmentally sustainable practices. It is hard, for example, to waste quality drinking water when only rainwater is available for cooking and sinks, drinking water has to be manually collected, and there is no toilet to flush it down (Figs. 14.2a, b). In other words, it is easier to be ecological than not. Likewise, because

Fig. 14.2a and 14.2b Green Hills spring water and rainwater systems

Fig. 14.2a and 14.2b (continued)

they have an uneven electricity supply, especially in winter, there are limits to what can be powered. Will and May have limited the number of phone chargers so that mobile phones deliberately have to be rotated to be charged. This sometimes causes tensions in the family, especially between the teenage children, but it also enforces the notion of limits. While these limits are hard to transgress, they do have unintended consequences. Domestic tasks take longer because of a lack of convenient infrastructure, and in particular these limits mean that the family does not have a fridge because there is not enough electricity to power it. A lack of fridge means they have to go to the shop more often, normally by car, to buy perishable goods. Another family on site has a more sophisticated house, with spring water piped into the kitchen sink, and more space and facilities. When May was asked what she would like, she answered, *"a fridge, a washing machine, a bathroom"* (May, Green Hills, interview). In time the expectations are likely to ratchet up rather than stay stable. In all, Green Hills has built new society-environment relations of limiting impact on the environment, but over time the impacts of consumption are slowly increasing. The very slow temporalities of change embedded in eco-communities mean that over time there is a quest to drop this self-provisioning when alternative easier options become available. Labour 'costs', skills and capacity are rarely considered but this lifestyle and the site itself are not accessible to those with differently abled bodies.

The experience of Green Hills suggests that while eco-communities demonstrate the possibilities of self-built off-grid infrastructures and that these infrastructures can dramatically reduce carbon emissions, they can also embed paradoxes which ultimately counter the positive gains made. Any self-provision is dynamic, not fixed, and without careful attention has the potential to morph into more carbon-intensive forms. In other words, these infrastructures need to remain shared collective endeavours, rather than individualized, to offer the greatest transformative potential.

Building New Social Infrastructures: The Case of LILAC

LILAC (Low Impact Living Affordable Community) is a 20-household cohousing eco-community in West Leeds, UK. It is a good example through which to explore how new social relations of sharing, through their social infrastructures, enable a reduction of environmental impact (Chatterton 2013) as shown in Fig. 14.3.

LILAC is a dense urban development and therefore not surprisingly homes have been designed to share common energy infrastructure (main connection and photovoltaic panels) and a Sustainable Drainage System (SUDS) wastewater management system. But residents also share gardens, bike sheds, laundry, car parks, a common house and spare bedrooms located in some blocks' hallways. The sharing of a laundry (and a contract preventing residents having their own washing machines), common house and the shared four guest rooms have enabled the individual houses to be smaller. The structural design of the site and homes reduces the overall environmental impact. This design also influenced other daily practices of households. For example, residents share what jobs they would like help with, or ask to borrow a piece of equipment use a WhatsApp group. Alan describes how daily tasks get shared:

> *There's a lot of efficiency of cohousing, of sharing errands. Frequently people say oh I'm just nipping to the supermarket, does anybody want anything? There's a lot of efficiencies of time use and so energy as well. And with that, with informal and formal child care as well, and then also sharing of tools and resources of bikes and tools and tents and many things. So that makes life easier and can have a better standard of living really with better stuff, because we share them and have a bit more time—which I hadn't appreciated how extensive that would be before we moved in.* (Alan, LILAC, interview)

> *There are weekly communal meals and residents also share each other's houses when guests visit: One of the things that we didn't even talk about and didn't anticipate was how much, when we're away, we'd lend each other our houses or flats for when people come and visit.* (Alan, LILAC, interview)

In order to maintain these shared spaces, there are team task work groups. Resident contributions to this communal work are uneven and those who work full time off-site claim they do not have time. As Fran, a co-founder, says, "it's possible that we made a mistake from the beginning. Perhaps only people who work part-time or less can be part of it" (LILAC, interview 2016). While LILAC has actively sought to create new social relations, which have implications for society-environment relations, there are residents who have struggled with this new sociality and sharing. Fran decided after two years that she could no longer cope with sharing:

> *I love the flat, I love parts of LILAC, but some of the behaviour of people drives me bonkers. One of them that comes to mind is soap in the soap dispenser in the washing machine. And how most people would put it in carefully into the correct bit, people here throw it in, because it's all over the floor, it's in the wrong things … and it will just sit there and it'll go in mine as well as theirs … Sometimes I think just get over it, it's just a bit of soap, and other times I want to throw it at somebody.* (Fran, LILAC, interview)

Fig. 14.3 LILAC physical and social infrastructures

In terms of accessibility and affordability, the ethic of sharing at LILAC extends beyond the design of the homes themselves to include the way in which the project was funded to be affordable. LILAC uses a solidaristic funding model (a Mutual Home Ownership Model) where the wealthier residents basically subside those less well-off. All residents only pay 35% of their annual income to live at LILAC and, therefore, the overall costs are shared unevenly between the community. This was because, as Alan put it, *"none of us wanted to live with just rich people"* (Alan, LILAC, Interview 2016). Overall, there is an interdependence of social and physical infrastructures—they have been purposefully designed to share. But there are also limits of who gets access to these infrastructures—much of LILAC is private despite earlier promises of being publicly shared.

14.5 Tensions of Self-Built Infrastructures

There are interesting tensions that emerge through this analysis of these self-built infrastructures. These infrastructures sought to both *facilitate and enforce* environmental sustainability. Yet, there is a tension in residents pushing back against the limits they have designed themselves—be that sharing white goods or a quest for a fridge. Is this an unsolvable tension, or is there a point at which residents accept these infrastructural limits? Furthermore, these self-built infrastructures have taken *considerable time* to create and, of course, use. Yet we have little time in which to invest in infrastructures that are socially and environmentally sustainable. There is an urgency to this need which these examples do not necessarily help us resolve, and the extent to which they can emerge as a component of broader, more strategic forms of climate urbanism, remains to be explored. There is also a relation between the temporalities of using these self-built infrastructures, *social reproduction and gender*, expressed in the embodied and variable experiences of everyday life in these eco-communities. It was women who bore most of the burden of the additional work and it was women who articulated a desire for more convenient infrastructures (see also Pickerill 2015a). There are many *embedded paradoxes in eco-communities*, such as "not having a fridge which increases the amount of food miles to go shopping" which residents acknowledge but either they disagree on its importance (this is often gendered), feel unable to change, or see it as temporary (when actually it is often long-standing). Despite a quest to self-build infrastructure in part to ensure autonomy in a changing climate, it is unclear how these infrastructures facilitate adaptation to climate change. They enable self-provision, but they are also fragile and at risk of water shortages or temperature extremes. The question of what more flexible infrastructures might look like, therefore, remains unclear. Is it the small scale on which these infrastructures have been built that will enable their continued adaptation, the low skills required to build and operate them, their collectivity and/or their low cost?

14.6 Conclusions

These two examples highlight the potential implications of eco-communities for climate urbanism. First, we must examine the unintended consequences of self-built physical infrastructures. They can be used to enforce limits, but this can lead to energy and time use elsewhere. Second, there are ratcheting demands of infrastructure. Residents sought more and would abandon self-provision if there were more comfortable alternatives available. Therefore, infrastructures need to be flexible and dynamic to changing demand. For example, if certain social infrastructures are built-in by physical design and not flexible to change, then it cannot accommodate tensions in use and simply leads to abandonment. Third, the temporalities of building, maintaining or relying on self-built physical infrastructures are rarely considered in eco-communities. Self-provision takes considerably longer, yet a cost-benefit analysis is not done. The social infrastructure of sharing is perceived to save time and be efficient, but to be part of this, residents need to have time to invest in being part of community activities.

Self-provided infrastructures in eco-communities are important forms of collective experimentation (open, creative, learning by doing) to seek positive change. Eco-communities are more than defensive eco-enclaves and engender new forms of collectivity despite these experiments being messy, unfinished, fluid and open. There remains a need to be critically cautious of how privileged many in eco-communities are, especially when suggesting urban dwellers can easily replicate these infrastructural forms. But it is this unfinished and open nature that we can learn most from for a new climate urbanism: it shows that it is necessary and worthwhile to engage in seemingly small-scale collective change attempts despite the final outcomes being unclear. It is in the very process of grassroots collective (social as much as material) experimentation that the possibilities and difficulties of transformation are realized. It is through such examples that the notion that a single policy or technology will suffice in response to climate change is once again challenged. Instead, we need diversity and risk-taking in climate urbanisms.

References

Anderson, J. (2017). Retreat or re-connect: How effective can ecosophical communities be in transforming the mainstream? *Geografiska Annaler B: Human Geography, 99*(2), 192–206.

Argüelles, L., Anguelovski, I., & Dinnie, E. (2017). Power and privilege in alternative civic practices: Examining imaginaries of change and embedded rationalities in community economies. *Geoforum, 86*, 30–41.

Bhakta, A., & Pickerill, J. (2016). Making space for disability in eco-housing and eco-communities. *The Geographical Journal, 182*(4), 406–417.

Caprotti, F., & Cowley, R. (2017). Interrogating urban experiments. *Urban Geography, 38*(9), 1441–1450.

Cattaneo, C. (2015). Eco-communities. In G. D'Alisa, F. Demaria, & G. Kallis (Eds.), *Degrowth: A vocabulary for a new era* (pp. 165–168). London: Routledge.

Chatterton, P. (2013). Towards an agenda for post-carbon cities: Lessons from LILAC, the UK's first ecological, affordable, cohousing community. *International Journal for Urban and Regional Research, 37*(5), 1654–1674.

Chitewere, T. (2018). *Sustainable community and green lifestyles*. London: Routledge Press.

Davies, G. (2010). Where do experiments end? *Geoforum, 41*, 667–670.

Ergas, C. (2010). A model of sustainable living: Collective identity in an urban ecovillage. *Organization and Environment, 23*(1), 32–54.

Fois, F. (2019). Enacting experimental alternative spaces. *Antipode, 51*(1), 107–128.

Hodson, M., & Marvin, S. (2010). Urbanism in the anthropocene: Ecological urbanism or premium ecological enclaves? *City, 14*(3), 298–313.

Holling, C. S., & Gunderson, L. H. (2002). Resilience and adaptive cycles. In: Panarchy: Understanding Transformations in Human and Natural Systems, 25–62.

Jarvis, H. (2019). Sharing, togetherness and intentional degrowth. *Progress in Human Geography, 43*(2), 256–275.

Lawhon, M., Nilsson, D., Silver, J., Ernstson, H., & Lwasa, S. (2018). Thinking through heterogeneous infrastructure configurations. *Urban Studies, 55*(4), 720–732.

Pascoe, B. (2014). *Dark Emu: Black seeds: Agriculture or accident?* Broome: Magabala Books.

Pickerill, J. (2015a). Bodies, building and bricks: Women architects and builders in eight eco-communities in Argentina, Britain, Spain, Thailand and USA. *Gender, Place and Culture, 22*(7), 901–919.
Pickerill, J. (2015b). Building the commons in eco-communities. In S. Kirwan, L. Dawney, & J. Brigstocke (Eds.), *Space, power and the making of the commons*. London: Routledge.
Pickerill, J. (2016). *Eco-homes: People, place and politics*. London: Zed Books.
Pickerill, J. (2019). Experimentations, in Antipode Editorial Collective (Eds.) Keywords in Radical Geography: Antipode at 50. Wiley.
Pickerill, J., & Chatterton, P. (2006). Notes towards autonomous geographies: Creation, resistance and self-management as survival tactics. *Progress in human geography, 30*(6), 730–746.
Pieterse, E. (2008). *City futures: Confronting the crisis of urban development*. London: Zed Books.
Powell, R. C., & Vasudevan, A. (2007). Geographies of experiment. *Environment and Planning A, 39*, 1790–1793.
Rice, J. L., Cohen, D. A., Long, J., & Jurjevich, J. R. (2020). Contradictions of the climate-friendly city: New perspectives on eco-gentrification and housing justice. *International Journal of Urban and Regional Research, 44*(1), 145–165.
Silver, J. (2018). Suffocating cities: Urban political ecology and climate change as social-ecological violence. Urban political ecology in the anthropo-obscene: Interruptions and possibilities, 129–146.

15

Conclusions: Three Modalities for a New Climate Urbanism

Vanesa Castán Broto, Enora Robin, and Aidan While

15.1 Introduction

As climate change transforms our societies, it also changes how we interact with our surroundings. Nowhere is this more evident than in urban areas. Climate change is slowly shifting our perceptions of urban safety and risk following the experiences of extreme weather events such as cyclones, hurricanes, storms, suburban wildfires, flooding and heatwaves. Climate migration is now affecting cities around the world. Thus, climate insecurity has shifted from being a concern of 'exceptional' places in the Global South to a concern for cities everywhere. If there were any doubt that cities have entered a new phase of environmental insecurity, the COVID-19 pandemic has thrown into sharp focus the fragility of protective infrastructures and prevailing economic systems, and the scale of our exposure to new biological threats. Public health has again become a priority for cities, but in new ways that consider their integration into

V. Castán Broto (✉) • E. Robin • A. While
Urban Institute, University of Sheffield, Sheffield, UK
e-mail: v.castanbroto@sheffield.ac.uk; e.robin@sheffield.ac.uk; a.h.while@sheffield.ac.uk

V. Castán Broto et al. (eds.), *Climate Urbanism*,
https://doi.org/10.1007/978-3-030-53386-1_15

multiple circuits, from local personal contacts to global population movements. At the same time, the need to provide a coherent response to climate change and reduce our dependence on fossil fuels entails the reorganization of the built environment with new concepts of mobility, thermal comfort, economic relations and communications to meet the aim of reducing carbon emissions. As shown throughout this book, the manifestations of this New Climate Urbanism are diverse. They reflect multiple geographies and varied biophysical contexts that differentiate climate change responses across countries, regions, and neighbourhoods. However, there is a common trend that is evident across all chapters: the New Climate Urbanism represents a qualitative shift in the way we look at cities in terms of safety, resilience, sustainability and broader development trajectories. The prevailing business-as-usual is gone.

If we accept that the New Climate Urbanism is here to stay, the question is how to ensure that future cities are able to adapt in ways that are inclusive and fair, addressing the needs of the most marginalized and preventing the emergence of spaces of exclusion as manifested in the creation of securitized and 'green' urban enclaves. The Sustainable Development Goal 11 aims to achieve cities that are inclusive, safe, resilient, and sustainable (United Nations 2015). Of these terms, the last three are often considered in the kinds of actions that lie at the core of climate urbanism. However, as the chapters in this book illustrate, climate action in cities can reinforce existing urban social inequalities and exclusions if it does not explicitly aim to achieve social justice.

The various chapters have explored what the New Climate Urbanism might mean in different places and around different domains of climate policy. These contributions highlight the renewed urgency for action in a context where climate displacement, climate poverty, and climate deaths are increasing year on year in the Global North and in the Global South. They show that we need to think carefully about the new forms of environmental injustices generated under the shadow of urban climate policy. The notion of 'climate apartheid' is increasingly used (Chap. 3) to capture the sense of a world divided by differential exposures to climate risks and differential access to the resources needed to protect the lives and livelihoods that are under threat.

Moreover, this New Climate Urbanism emerges against the backdrop of other phenomena changing the context of action. Climate migration and displacements will undoubtedly contribute to intensify the tendency towards global urbanization. Climate change will worsen crises of food and water supply and will force us to build more resilient systems of food, energy, housing and water provision. COVID-19 will not only refocus attention on urban public health but will also support more intensive behavioural control. It will perhaps lead us to rethink the design, layout, density and interconnectedness of cities. The search for liveable places will continue, dividing urban areas between the affluent and the dispossessed. Affluence alone, however, will not be sufficient to protect cities from the effects of climate change. Some of today's wealthier cities might decline as wealthier citizens and economic activities gravitate towards less vulnerable locations or to cities that can manage climate stress through technology (e.g. controlled environments and climatic modification, Marvin and Rutherford 2018).

This book thus constitutes an initial exploration of what this New Climate Urbanism looks like, how it manifests, and what its consequences might be. Whether the New Climate Urbanism promotes actions that work for all is the central question that inspires the contributions of this book. Examples abound about new environmental injustices as cities selectively adapt to climate change and there is a suspicion that the New Climate Urbanism will replicate and perhaps reinforce the prevailing injustices of the twenty-first-century city. There is therefore an urgent and compelling need to explore the potential of New Climate Urbanism to create opportunities for a sustainable and just city, including the potential to link climate change, public health and migration as part of a new humanitarian perspective on cities. The following sections explore some of the expanding areas for a research agenda on the New Climate Urbanism moving forward.

15.2 Rethinking Multiple Pathways to the Future

Take a moment to close your eyes and envision the world, say, 10 or 15 years from now. What you're imagining, it's quite likely, is a lot of new technology. In general, I've found that when we consider major world problems like poverty, climate change or cancer, we optimistically think about a techno-utopia that solves them. There's nothing wrong with that, but we have to move away from looking at the future in just this one way. I do everything in my power not to talk about a single future but to talk about futures instead. Open yourself up to considering all kinds of possible scenarios and all kinds of solutions. (Ari Wallach 2017)

No one single technological future will solve the complex questions that we are facing, Wallach explains. Thus, we need to engage with a multiplicity of futures. In a seminal paper on the future of the energy transition in the UK, Rydin et al. (2013) argued for the deployment of a coevolutionary perspective to study possible transitions within existing infrastructure systems. They wanted us to imagine transitions as more than "*a discrete series of changes in technologies and associated infrastructure*" to examine them instead

as the outcome of on-going interactions between technologies, political and economic frameworks, and between institutions and social practices, during which these different dimensions change or co-evolve together to produce distinct pathways of change. (Rydin et al. 2013)

This is a perspective that has greater purchase in sustainable development and environmental politics (e.g. Leach and Scoones 2010) than in urban planning. Opening rather than closing our imagination to multiple future pathways is imperative to sustain urban life in a climate-changed world. Such analyses resonate with futurists' work, whose role is to open up our ability to imagine alternatives beyond techno-utopias (see Wallach's quote above, and Hajer and Pelzer 2018). The closing down of the human imagination in the face of impending disaster seems to be as restrictive as comprehensive planning schemes were for the planners and urban managers of the past. Castán Broto and Westman (2019: 2) have argued in relation to urban climate futures, "*there is no single vision of that future. Imposing such global visions in one single locality would be akin to impose ill-fitting solutions to problems that may not exist.*"

Paradoxically, our imagination seems to fail us precisely in the search for utopia, a state of seemingly unending possibilities. This book has analysed both the search for imaginative responses to climate change and existing attempts to settle debates on a fixed range of models that will define urban futures. With hyperbolic claims about how they can make or break climate futures, cities have raised expectations with regards to their ability to transform themselves in response to the climate crisis. These expectations may well be beyond current possibilities, but attempts to address climate change in cities do exist, and they take various forms. Many are motivated by a 'climate opportunism' that fails to address the root causes of the climate crisis and instead perpetuates narratives of endless growth fuelled by resource extraction. In doing so, these approaches reinforce intersecting economic, social, ethnic, gender, racial and spatial inequalities. But those are not the only responses we see emerging. In the next section, we highlight three different modalities of climate urbanism that have emerged throughout this edited book. We have called them 'reactive,' 'entrepreneurial,' and 'transformative'; they coexist and at times contradict each other.

15.3 Three Modalities of a New Climate Urbanism: Entrepreneurial, Reactive, Transformative

Looking into current and future developments, it is possible to envisage that the New Climate Urbanism will contribute to legitimizing authoritarian forms of governance to expand and protect the capitalist economic functions of cities. This 'entrepreneurial' approach to climate urbanism raises critical questions for future research, including how current decarbonization and climate adaptation pathways in cities reproduce neoliberal and resource-intensive forms of urban development. Linda Shi

(Chap. 4) stresses how current analyses of 'climate urbanism' actually identify the legacy of logics of uneven urban development, resource extraction, and climate gentrification that emerged before climate change became a challenge of urban policy. This competitive urban protectionism is consistent with ideas of urban entrepreneurialism and neoliberal spatial competition, now imbued with a new climatic dimension. 'Entrepreneurial' climate urbanism is associated with the deployment of neoliberal principles of urban management to address climate change in urban areas. Indeed, current approaches to climate action in cities show that various interests, from local governments to IT companies, real estate actors, businesses, or development banks, see climate change as an economic opportunity. Entrepreneurial climate urbanism sees climate action as an opportunity to spur 'green' economic growth, as shown by Corina McKendry's analysis of Colorado Springs climate-friendly policies (Chap. 9). In the literature, climate gentrification is a growing concern that comes to add to parallel analyses of how green policies are creating new forms of exclusion and oppression in contemporary cities (Anguelovski et al. 2018). A focus on climate-friendly 'entrepreneurial' strategies can help climate urbanism research to challenge approaches that dilute climate action within 'business as usual' urban boosterism.

However, much of the climate urbanism that has emerged so far is not purposively entrepreneurial, but rather it is reactive, characterized by the defence of the existing economic status quo within the narrow horizons of economic competitiveness. As Joshua Long, Jennifer Rice and Anthony Levenda (Chap. 3) demonstrate, this reactive approach to protecting core economic functions is not socially neutral. It is leading to new forms of 'climate apartheid' as wealthy property interests are protected at the expense of the less affluent. In Chap. 8, Eric Chu explored how the concept of resilience has gained traction as a way to direct economic resources into evidence-based planning interventions in two Indian cities. His analysis shows how resource constraints in cities make them more dependent on short-lived, external funding to implement climate action, limiting its transformative potential and relying on technocratic-led processes that favour the protection of economic activities in the short run. In Chap. 9, Marta Olazabal shows that a wide range of adaptation plans around the world do not have the evidence on climate risks needed for robust planning. Lack of evidence hinders the ability of local governments—and that of other actors—to anticipate and react to climate change impacts over different timescales, locking cities into short-term reactive responses. Sustainable and socially inclusive approaches to climate change are unlikely to emerge from reactive responses to climate stresses and shocks in cities. Climate change requires distinct ways of thinking about urban management and progressive pathways in order to protect the vulnerable and dispossessed.

This book explored what more socially 'transformative' approaches to climate urbanism could look like. Socially transformative climate urbanism relates to the growing efforts by multiple actors and citizens to align climate action within a broader framework of empowerment, social protection and resource redistribution to reshape urban societies' relationship with nature (Enora Robin, Linda Westman and Vanesa Castán Broto, Chap. 2). Most research on urban climate policy to date has focused on 'the Global North' and the capacity of local states to act and to leverage resources for climate action. Expanding the geography of climate urbanism research brings a wider focus on other forms of governance that are more fragmented and complex as discussed by Sikku Juhola in Chap. 5. Furthermore, in Chap. 12, Andrew P. Kythreotis and Theresa Mercer explore the potential of new forms of collaborative educational strategies to support intergenerational learning and empowerment. This raises questions as to how future research on climate urbanism can be integrated into pedagogical practices that spur collective action and intergenerational learning in order to transform cities. But the manners of transformation are highly variable. For example, this book discusses the role of community-driven projects in the production of socially just and transformative climate urbanism. In Chap. 13, Long Seng To discusses the example of community projects in Nepal and Malawi as innovative forms of governance to support adaptation to climate risks, building on local knowledge(s) in the context of decentralized governance. The question of transformation, in future climate urbanism research, should thus attend to the capacity of non-state actors to spark change. Jenny Pickerill in Chap. 14 reflects upon her long experience of studying eco-communities, showing how these collective modes of organizing housing and infrastructure provision provide opportunities to change broader cultures, our relationships to resource limits, nature and society. She also stresses that eco-communities can only be part of a much broader range of actions for rapid and large-scale transformation. Indeed, a key challenge for climate urbanism research will be its capacity to reframe the relationship between 'the urban' and climate change in a way that can spark transformative action at different scales of governance. In Chap. 6, Westman and Castán Broto start to explore these issues by looking at whether and how urban imaginaries are integrated into international climate policy. In Chap. 7, James J. Patterson brings to the fore the complexities of institutional change when thinking about transformative action. In doing so, he stresses the need for future research on climate urbanism to combine prescription (what should be done in principle) with a better understanding of processes of change, remaking the rules-in-use, and understanding the consequences of deliberate action within different institutional and historical contexts. In Chap. 11, Luna Khirfan highlights the transformative potential of urban design thinking in understanding and shaping urban responses to climate change, showing how the micro-scale of urban planning impacts directly on the health and well-being of citizens.

The New Climate Urbanism constitutes a new mode of being for urban areas, a mode of being that forces urban actors to connect to a wider global context of environmental change. The different contributions presented in this book offer initial reactions to this qualitative change as a new way of thinking urban life, and consider its implications for the reproduction of urban life. The three modalities of climate urbanism discussed here—entrepreneurial, reactive, and socially transformative—reflect the deep contradictions at the core of actions making up this New Climate Urbanism.

15.4 Conclusion

Inherent to the New Climate Urbanism is the necessity of a global urban climate perspective that recognizes the differential challenges and capacities of cities to respond and adapt to a climate-changed world. As ever with climate change, acting now to support cities to adapt will help mitigate some of the future impacts of climate change, and that includes the strategic management of urban growth and decline, displacement and migration. The book has highlighted powerful tendencies for climate change and climate action in cities to reinforce dynamics of exclusion. However, there are also opportunities for governments, citizens and organizations to think differently about the design and management of cities and the distribution of resources in a finite world. All the contributions of this book were written before the COVID-19 pandemic of 2019–2020 but the key messages of the book resonate with the changed perspectives brought about by the coronavirus. Both COVID-19 and climate change highlight the importance of coordinated action at national and local scales, but also the need to move beyond the laissez-faire market-led approach to social protection. COVID-19 became a global crisis because governments had failed to hear the warnings of a threat to human life that had long been recognized and predicted. COVID-19 discriminates by health and living conditions, environmental factors and differential access to resources and infrastructures that could help to maintain human life and livelihoods. Climate change will unfold over a longer period of time than COVID-19 but in the post-COVID-19 landscape we have an opportunity to rethink social life and urban futures in ways that integrate climate change and a deeper concern for human-nature relations.

An emerging theme that will no doubt find echo in future research relates to the drivers of change that can help mainstream climate change mitigation and adaptation strategies across sectors, stakeholders, and scales of governance. Will the New Climate Urbanism consolidate into a homogeneous model, or will it create space for alternative models to adapt to the changing needs of urban areas and their citizens? A central concern emerging from the dialogue among the scholars in this book is whether the New Climate Urbanism will empower vulnerable communities to partake in the design of future climate action within and across cities. Can anyone deliver alternative models of climate urbanism that include vulnerable groups and that address the entrenched inequalities that create vulnerabilities in the first place? Alternative modes of thinking and acting, perhaps grounded in feminist and postcolonial theories, decolonial praxis, and other deliberate attempts to think beyond the mainstream, may help us to move beyond a critique of neoliberalism to understand processes of radical and incremental change, and to explore how climate action can support the emergence of new urban worlds in the age of climate change.

References

Anguelovski, I., Connolly, J., & Brand, A. L. (2018). From landscapes of utopia to the margins of the green urban life: For whom is the new green city? *City, 22*, 417–436.

Castán Broto, V. C., & Westman, L. (2019). *Urban sustainability and justice: Just sustainabilities and environmental planning*. London: Zed Books Ltd.

Hajer, M. A., & Pelzer, P. (2018). 2050—An Energetic Odyssey: Understanding' Techniques of Futuring' in the transition towards renewable energy. *Energy research and social science, 44*, 222–231.

Leach, M., Scoones, I., & Stirling, A. (2010). *Dynamic sustainabilities: Technology, environment, social justice*. London: Routledge.

Marvin, S., & Rutherford, J. (2018). Controlled environments: An urban research agenda on microclimatic enclosure. *Urban Studies, 55*(6), 1143–1162.

Rydin, Y., Turcu, C., Guy, S., & Austin, P. (2013). Mapping the coevolution of urban energy systems: Pathways of change. *Environment and Planning A, 45*(3), 634–649.

United Nations. (2015). 2030 Agenda for Sustainable Development. Retrieved from https://www.un.org/ga/search/view_doc.asp?symbol=A/RES/70/1& Lang=E.

Wallach, A. (2017). How to think like a futurist. Ideas.ted.com.